Developments in
Characteristic Function Theory

Griffin books of cognate interest

Characteristic functions   E. LUKACS

This monograph, much extended in its second edition, establishes the basic theory of characteristic functions, which have a fundamental place in probability and mathematical statistics. Detailed proofs are given, including important results appearing in the foreign literature.

Applications of characteristic functions   E. LUKACS and R. G. LAHA

    Discusses applications of characteristic functions to mathematical statistics, using for the most part the methods of analytic probability theory.

The advanced theory of statistics
        SIR MAURICE KENDALL, A. STUART and J. K. ORD

    Vol. 1 of this comprehensive work has a 30-page chapter on characteristic functions with a discussion of some applications.

Green's function methods in probability theory   J. KEILSON

Families of bivariate distributions   K. V. MARDIA

Families of frequency distributions   J. K. ORD

An introduction to infinitely many variates   E. A. ROBINSON

Developments in Characteristic Function Theory

EUGENE LUKACS

The Catholic University of America
Washington, D.C.

MACMILLAN PUBLISHING CO., INC.
NEW YORK

Copyright © Eugene Lukacs 1983

Published in USA by
Macmillan Publishing Co., Inc.
866 Third Avenue, New York, N.Y. 10022

Distributed in Canada by
Collier Macmillan Canada, Ltd.

All rights reserved. No part of this book may be reproduced or transmitted in any form or by any means, electronic or mechanical, including photocopying, recording, or by any information storage and retrieval system, without permission in writing from Charles Griffin & Co. Ltd., Charles Griffin House, Crendon Street, High Wycombe, Bucks, England, HP13 6LE.

By arrangement with the originating publisher
CHARLES GRIFFIN & COMPANY LIMITED
London and High Wycombe

First published 1983

Library of Congress Catalog Card Number: 83-61216
ISBN: 0-02-848550-5

Printed in Great Britain

Preface

Characteristic functions were originally introduced as tools for the study of limit theorems of probability theory. Later it became clear that the theory of characteristic functions is also of intrinsic mathematical interest. The author's books on characteristic functions (published in 1960 and 1970) were primarily motivated by a wish to study the mathematical theory of characteristic functions. A good deal of work has been done since the publication of the 1970 edition, and the theory has been advanced in a very significant way.

The present monograph is intended to cover these later results. The first chapter gives a brief summary of the basic concepts. These are presented without proofs, but with adequate references to the book published in 1970. Proofs are given only for Theorems 1.2.3 and 1.2.4 since these are more recent results, not contained in the earlier monograph. Chapters 2 to 11 cover the recent advances in the theory of characteristic functions, while the last four deal with closely related subjects. Chapter 12 treats the recently developed theory of analytic distribution functions, and Chapter 13 discusses metrics in the space of distribution functions. The final two chapters are concerned with some generalizations of characteristic functions, namely with ridge functions and functions of bounded variation.

The author is grateful to Professors S. J. Wolfe, T. O'Connor, V. K. Rohatgi, H. J. Rossberg, G. Laue, M. Dewess, G. Siegel and B. Jesiak for comments on certain parts of the manuscript. Thanks are also due to the Technical University of Vienna and to the University of Erlangen Nürnberg for helping to procure the pertinent literature and for arranging for the typing of the manuscript, although part of it was typed at the Catholic University of America.

EUGENE LUKACS

Washington D.C.
1982

Contents

Preface		v
1 INTRODUCTION		1
1.1	Distribution functions and characteristic functions	1
1.2	Conditions on characteristic functions	4
1.3	Infinite divisibility	12
1.4	Analytic characteristic functions	14
2 MOMENTS OF ARBITRARY POSITIVE ORDER		17
2.1	Existence of moments; behaviour of distribution functions and characteristic functions	17
2.2	Expansion of characteristic functions	21
2.3	Fractional moments and fractional derivatives	25
3 ESTIMATION OF THE CLOSENESS OF DISTRIBUTION FUNCTIONS		31
3.1	Distance definitions	31
3.2	Esseen's results and their extension	33
3.3	Distances in the Lévy metric	38
4 EXTENSIONS OF THE CONCEPT OF STABLE DISTRIBUTIONS		45
4.1	Generalizations of stable distributions	45
4.2	Self-decomposable distributions	47
5 UNIMODALITY		49
5.1	Conditions for unimodality	49
5.2	Convolutions of unimodal distributions	50
5.3	Strong unimodality	53
5.4	Unimodality of stable and self-decomposable distributions	57
6 FACTORIZATIONS AND INFINITE DIVISIBILITY		64
6.1	Some properties of absolutely continuous or discrete infinitely divisible distributions	64
6.2	Tail behaviour of infinitely divisible distributions	68
6.3	Decomposition of symmetric stable distributions	70
6.4	Certain indecomposable distributions	71
7 ANALYTIC CHARACTERISTIC FUNCTIONS		76
7.1	Analytic continuation of analytic characteristic functions	76
7.2	Entire characteristic functions satisfying certain conditions	80
7.3	Entire characteristic functions of order 2 with a finite number of zeros	86
7.4	Special families of entire and analytic characteristic functions	92
7.5	Convolution of Poisson type distributions	101
7.6	I_0 contains convolutions of Poisson type distributions	103
8 INFINITELY DIVISIBLE DISTRIBUTIONS DETERMINED BY THEIR VALUES ON A HALF-AXIS		105
8.1	The case of the normal distribution	105
8.2	Generalizations of Theorem 8.1.1	110

CONTENTS

9 A GENERALIZATION OF THE DECOMPOSITION PROBLEM 112
   9.1 α-decompositions 112
   9.2 The class I_0^α 114

10 BOUNDARY CHARACTERISTIC FUNCTIONS 116

11 MIXTURES OF DISTRIBUTION FUNCTIONS 119
   11.1 Infinite divisibility of mixtures 119
   11.2 Mixtures of exponential distributions which are not infinitely divisible 121

12 ANALYTIC DISTRIBUTION FUNCTIONS 123
   12.1 Definition of analytic distribution functions and some of their properties 123
   12.2 Continuation of distribution functions 127
   12.3 Limit theorems and restricted convergence 128

13 METRICS IN THE SPACE OF DISTRIBUTION FUNCTIONS 131
   13.1 Properties of metrics defined on a set X; comparison of metrics 131
   13.2 The case where X is the set of distribution functions 134
   13.3 Metrics on the set of characteristic functions 135

14 RIDGE FUNCTIONS 139
   14.1 Definition of ridge functions and the existence of ridge functions which are not analytic characteristic functions .. 139
   14.2 Elementary properties of ridge functions 141
   14.3 Factorization of ridge functions 143
   14.4 Entire ridge functions 147

15 FUNCTIONS OF BOUNDED VARIATION 150
   15.1 Fourier–Stieltjes transforms of functions of bounded variation 151
   15.2 Decomposition of functions of bounded variation 153

APPENDIX 164

LIST OF EXAMPLES 169

REFERENCES 170

INDEX 179

1 Introduction

In this chapter we summarize some of the definitions and theorems concerning distribution functions and characteristic functions which will be used in the sequel. Most of these statements are given without proof; the proofs and additional results can be found in the monograph by E. Lukacs (1970).

1.1 Distribution functions and characteristic functions

A function $F(x)$ is said to be a distribution function if it is non-decreasing, continuous to the right, and if $F(+\infty) = 1$, $F(-\infty) = 0$.(*)

A point x is called a discontinuity point of a distribution function $F(x)$ if $F(x) = F(x+0) \neq F(x-0)$. If $F(x) = F(x-0)$ then we say that x is a continuity point.

Theorem 1.1.1. A distribution function has at most a denumerable set of discontinuity points.

Theorem 1.1.2. Let $F(x)$ be a distribution function. Then it can be decomposed in the following way:

$$F(x) = a_1 F_d(x) + a_2 F_{ac}(x) + a_3 F_s(x).$$

Here $a_1 \geq 0$, $a_2 \geq 0$, $a_3 \geq 0$, $a_1 + a_2 + a_3 = 1$, $F_d(x)$ is a purely discrete distribution function, $F_{ac}(x)$ is absolutely continuous, and $F_s(x)$ is a purely singular distribution function.

A purely discrete distribution function can be written in the form

$$F(x) = \sum_j p_j \epsilon(x - \xi_j),$$

where $p_j \geq 0$, $\sum_j p_j = 1$ and where $\epsilon(x - \xi)$ is a degenerate distribution with a single saltus at the point $x = \xi$.

A (purely) absolutely continuous distribution function can be written in the form

$$F(x) = \int_{-\infty}^{x} p(y) \, dy,$$

(*) Here $F(+\infty) = \lim\limits_{x \to \infty} F(x)$, $F(-\infty) = \lim\limits_{x \to -\infty} F(x)$; similarly, $F(x+0) = \lim\limits_{h \downarrow 0} F(x+h)$, $F(x-0) = \lim\limits_{h \downarrow 0} F(x-h)$.

where $p(y)$ is called the frequency (or density) function of the distribution function $F(x)$.

Singular distribution functions are continuous but not absolutely continuous.

Let $F(x)$ be a distribution function and let k be a non-negative integer. If the integral

$$(1.1.1) \quad \alpha_k = \int_{-\infty}^{\infty} x^k \, dF(x)$$

exists, then it is called the (algebraic) moment of order k of $F(x)$. If the integral

$$(1.1.2) \quad \beta_k = \int_{-\infty}^{\infty} |x|^k \, dF(x)$$

exists, then it is called the absolute moment of order k of $F(x)$.

It can happen that a distribution function has only moments up to a certain order (or no moments of positive order). The absolute moment of order k exists if, and only if, the algebraic moment exists.

The following relations between moments are valid:

$$(1.1.3) \quad \begin{cases} \alpha_{2k-1} \leqslant |\alpha_{2k-1}| \leqslant \beta_{2k-1} \\ \alpha_{2k} = \beta_{2k} \\ \alpha_0 = \beta_0 = 1 \\ \beta_{k-1}^{1/(k-1)} \leqslant \beta_k^{1/k} \end{cases}$$

The last inequality was generalized by V. V. Petrov (1975)

$$\beta_r^{1/r} \leqslant \gamma^{(1/r)-(1/s)} \beta_s^{1/s} \quad (r<s),$$

where $\gamma = 1 - F(0) + F(-0)$.

Let $F(x)$ be a distribution function. Its Fourier-Stieltjes transform

$$(1.1.4) \quad f(t) = \int_{-\infty}^{\infty} e^{itx} \, dF(x)$$

is called the characteristic function of the distribution function $F(x)$.

We next list several important properties[*] of characteristic functions. If $f(t)$ is the characteristic function of a distribution function, then

- (a) $f(0) = 1, |f(t)| \leqslant 1, f(-t) = \overline{f(t)}$;
- (b) every characteristic function $f(t)$ is uniformly continuous in the real variable t;
- (c) Re $f(t)$ is also a characteristic function;
- (d) $|f(t)|^2 = f(t) f(-t)$ is a characteristic function.

Extensive tables of characteristic functions can be found in Oberhettinger (1973).

---

[*] The horizontal bar denotes the complex conjugate.

INTRODUCTION

Properties of characteristic functions and moments are closely connected. For these connections as well as for series expansions of characteristic functions we refer to Lukacs (1970).

Some of the important relations between characteristic functions and distribution functions are described by the following theorems.

Theorem 1.1.3 (Uniqueness theorem). *Two distribution functions are identical if, and only if, their characteristic functions are identical.*

Theorem 1.1.4 (Inversion theorem). *Suppose that $F(x)$ is a distribution function with characteristic function $f(t)$. Then*

$$F(a+h) - F(a) = \lim_{T \to \infty} \frac{1}{2\pi} \int_{-T}^{T} \frac{1-e^{-ith}}{it} e^{-ita} f(t) \, dt.$$

The convolution of two distribution functions F_1 and F_2 is defined as

$$F(z) = \int_{-\infty}^{\infty} F_1(z-x) \, dF_2(x) = \int_{-\infty}^{\infty} F_2(z-x) \, dF_1(x)$$

and we write frequently $F = F_1 * F_2$. Clearly F is a distribution function.

T. Kawata (1969) replaced the improper integral $\lim_{T \to \infty} \int_{-T}^{T}$ which occurs in Theorem 1.1.4 by the integral $\lim_{T, T' \to \infty} \int_{-T'}^{T}$ and gave the following necessary and sufficient condition for the existence of the asymmetric integral.

Corollary to Theorem 1.1.4. *In order that*

$$\lim_{T, T' \to \infty} \int_{-T'}^{T} \frac{1-e^{itx}}{it} f(t) \, dt$$

should exist, it is necessary and sufficient that

$$\lim_{\epsilon \to +0} \int_{\epsilon}^{1} \frac{G(u,x) - G(u,0)}{u} \, du$$

exists. Here $G(u, x) = F(u+x) - F(-u+x)$ and the limit equals $F(x) - F(0)$. (T and T' go independently to infinity.)

Theorem 1.1.5 (Convolution theorem). *A distribution function $F(x)$ is the convolution of two distribution functions $F_1(x)$ and $F_2(x)$, if and only if, the corresponding characteristic functions satisfy the relation $f(t) = f_1(t) f_2(t)$.*

It follows that the product of two characteristic functions is always a characteristic function.

We say that a sequence $\{F_n(x)\}$ of distribution functions converges weakly to a distribution function $F(x)$ if

$$\lim_{n\to\infty} F_n(x) = F(x)$$

for all continuity points x of $F(x)$. We write then

$$\mathrm{Lim}_{n\to\infty} F_n(x) = F(x).$$

Theorem 1.1.6 (*Continuity theorem*). *A sequence* $\{F_n(x)\}$ *of distribution functions converges weakly to a distribution function* $F(x)$ *if, and only if, the corresponding sequence of characteristic functions* $\{f_n(t)\}$ *converges to the characteristic function* $f(t)$ *of* $F(x)$ *at all points t.*

1.2 Conditions on characteristic functions

The following problem often arises:

Suppose that a function $f(t)$ of the real variable t is given. It is frequently necessary to decide whether $f(t)$ can be a characteristic function. In this section we deal with such criteria.

We first mention a necessary condition. Every characteristic function satisfies the relation

$$(1.2.1) \quad \mathrm{Re}\,[1 - f(t)] \geq \frac{1}{4^n}\,\mathrm{Re}\,[1 - f(2^n t)]$$

Other necessary conditions can be found in Lukacs (1970).

We present a necessary and sufficient condition which is due to S. Bochner (1932). In order to formulate this result we must introduce the concept of non-negative definiteness.

A non-negative definite function $f(t)$ is a complex-valued function $f(t)$ of the real variable t which is continuous and for which the sum

$$S = \sum_{j=1}^{N}\sum_{k=1}^{N} f(t_j - t_k)\,\xi_j\bar{\xi}_k$$

is real and non-negative for any positive integer N and any real $t_1, t_2, \ldots, t_N$ and complex $\xi_1, \xi_2, \ldots, \xi_N$.

Theorem 1.2.1. *The necessary and sufficient condition which a function* $f(t)$ *must satisfy in order to be a characteristic function is that* $f(t)$ *be non-negative definite and that* $f(0) = 1$.

An interesting and easily applicable condition is due to G. Pólya (1949).

Theorem 1.2.2. *A real-valued and continuous function* $f(t)$ *defined for all real t which satisfies the following conditions:*

INTRODUCTION

    (i) $f(0) = 1$
    (ii) $f(t) = f(-t)$
    (iii) $f(t)$ *is convex for* $t > 0$
    (iv) $\lim_{t \to \infty} f(t) = 0$

is always a characteristic function of an absolutely continuous distribution function $F(x)$.

Additional criteria for characteristic functions can be found in Lukacs (1970).

A particular case of Pólya's result is very convenient. If the function $f(t)$ is twice differentiable then one can replace condition (iii) of Theorem 1.2.2 by the requirement that

    (iii') $f''(t) \geq 0 \quad$ for $t > 0$.

In this case we easily obtain the following modification:

Corollary 1 to Theorem 1.2.2. Let $g(t)$ be a real-valued, even function which can be differentiated twice. Suppose further that $g(0) = 0$, $\lim_{t \to \infty} g(t) = -\infty$ and that $g''(t) + [g'(t)]^2 \geq 0$ for all $t \geq 0$. Then $f(t) = \exp[g(t)]$ is a characteristic function.

As a consequence one obtains the following statement:

Corollary 2 to Theorem 1.2.2. Let $h(t)$ be a real-valued, even function and suppose that the derivatives $h'(t)$ and $h''(t)$ exist and are bounded and that $h(0) = 0$ while $h(t) = O(t)$ as $t \to \infty$. Let $\lambda > 0$; then

$$f(t) = \exp\{-\lambda |t| + h(t)\}$$

is a characteristic function, provided λ is sufficiently large.

Corollary 2 follows immediately from corollary 1.

Corollary 3 to Theorem 1.2.2. Let $h(t)$ be a real-valued, bounded function which has bounded first and second derivatives. Let $\alpha > 0$ and let $g(t) = [-\lambda + h(\log|t|)]|t|^\alpha$. Then for sufficiently large $\lambda \geq 0$ the function $\exp[g(t)]$ is a characteristic function.

This statement follows easily from Corollary 1 to Theorem 1.2.2.
The inequality (1.2.1) can be generalized.

Theorem 1.2.3. Let n be a positive integer, then the inequality

$$1 - \operatorname{Re}[f(nt)] \leq n\{1 - [\operatorname{Re} f(t)]^n\} \leq n^2[1 - \operatorname{Re} f(t)]$$

is satisfied for every characteristic function $f(t)$.

For the proof we need the following lemma:

Lemma 1.2.1. *For all real x and all positive integers n one has $\cos^n x \leqslant (n-1+\cos nx)/n$.*

It is easy to show by induction that the inequality

$$(1.2.2) \quad \left| \frac{\sin nx}{\sin x} \right| \leqslant n$$

holds for all real x and all non-negative integers n. Let

$$(1.2.3) \quad f(x) = n \cos^n x - \cos nx$$

Then $f(x) \leqslant n$. Moreover,

$$f'(x) = -n^2 \cos^{n-1} x \sin x + n \sin nx.$$

Therefore $f(x)$ can have a (relative) maximum only if either

$$(1.2.4a) \quad \sin x = 0$$

or if

$$(1.2.4b) \quad \sin x \neq 0 \quad \text{and} \quad n \cos^{n-1} x = \frac{\sin nx}{\sin x}.$$

If $\sin x = 0$ then $x = k\pi$ (k integer) and it follows that $f(k\pi) \leqslant n-1$. If, however, (1.2.4b) holds, then

$$f(x) = \frac{\cos x \sin nx}{\sin x} - \cos nx$$

$$= \frac{\sin(n-1)x}{\sin x}$$

so that we see from (1.2.2) that

$$f(x) \leqslant n-1$$

and the lemma is proved.

For the proof of the theorem we have to distinguish two cases:

(a) n is an even integer. In this case x^n is a convex function. To simplify the notation we write $u(t) = \operatorname{Re} f(t) = \int_{-\infty}^{\infty} \cos tx \, dF(x)$ and we see from Jensen's inequality that

$$[u(t)]^n \leqslant \int_{-\infty}^{\infty} \cos^n tx \, dF(x)$$

INTRODUCTION

or, in view of Lemma 1.2.1,

$$n[u(t)]^n \le \int_{-\infty}^{\infty} (n-1 + \cos nx)\, dF(x)$$

or

$$nu^n(t) \le n - 1 + \int_{-\infty}^{\infty} \cos nx\, dF(x).$$

Hence

$$nu^n(t) \le (n-1) + u(nt).$$

This is the statement of the theorem. We still have to consider the case

(b) n is an odd integer. The statement is trivial if $n = 1$ or if $n[1 - u^n(t)] \ge 2$ and $n \ge 3$. Without loss of generality we can assume

$$n - 2 < nu^n(t).$$

Since $n \ge 3$ we have $1 \le nu^{n-1}(t)$ and $0 < u(t)$.

We consider the function

(1.2.5) $\quad y(x) = x^n - a^n - (x-a)na^{n-1}$

for $-1 \le x \le 1$ and $0 < a \le 1 \le na^{n-1}$ and $1 \le n - 2 < na^{n-1}$. An elementary computation shows that $y(x) \ge 0$ for $|x| \le 1$.

We put $a = u(t), x = \cos tx$ in (1.2.5) and see that

$$\cos^n tx \ge u^n(t) - [\cos tx - u(t)]\, nu^{n-1}(t)$$

We integrate this relation and get

$$\int_{-\infty}^{\infty} \cos^n tx\, dF(x) \ge u^n(t)$$

It follows then from Lemma 1.2.1 that

$$\int_{-\infty}^{\infty} [n - 1 + \cos ntx]\, dF(x) \ge nu^n(t)$$

or

$$1 - u(nt) \le n[1 - n^n(t)]$$

so that the theorem is proved.

Theorem 1.2.1 is due to C. R. Heathcote & J. W. Pitman (1972). In this paper the authors also derived several inequalities of a character similar to Theorem 1.2.2. Other inequalities for the absolute value of characteristic functions, which also involve moments, were given by H. Prawitz (1973), (1974).

Theorem 1.2.2 and Theorem 4.3.2 of Lukacs (1970) give sufficient conditions which assure that a real-valued function $f(t)$ of a real variable is a characteristic function. Next we give a related sufficient condition which concerns complex-valued functions of a real variable.

Theorem 1.2.4 (Shimizu). Let $\phi(t)$ and $\psi(t)$ be real-valued functions of the real variable t which satisfy the following conditions:

(i) $\phi(0) = 1$;
(ii) $1 \geqslant \phi(t) = \phi(-t) \geqslant 0$
$\psi(t) = -\psi(-t)$;
(iii) $\psi(t)$ *is differentiable and absolutely integrable over* $(0, \infty)$;
(iv) *the functions* $\phi(t) + c\psi'(t)$ *and* $\phi(t) + c \int_{-\infty}^{\infty} \psi(u)\, du$ *are convex on* $(0, \infty)$ *for all real c with* $|c| \leqslant 1$.

Then $f(t) = \phi(t) + i\psi(t)$ is a characteristic function. The distribution function F corresponding to f(t) has the form

$$F(x) = p\epsilon(x) + (1-p) G(x).$$

Here $\epsilon(x)$ is the degenerate distribution with discontinuity at $x = 0$, while $G(x)$ is an absolutely continuous distribution.

It follows from the assumptions of the theorem that $\phi(t)$ is monotonically non-increasing and that it is bounded from below by 0. Thus $\lim_{t \to \infty} \phi(t) = p$ exists and $1 \geqslant p \geqslant 0$. If $p = 1$, one sees from (ii) and (iv) that $f(t) \equiv 1$. We assume therefore that $p < 1$ and write $q = 1 - p$. We write also

(1.2.5a) $\quad h(t) = [\phi(t) - p]/q$,

so that

(1.2.5b) $\quad \phi(t) = p + qh(t)$.

The function $h(t)$ satisfies the conditions of Theorem 1.2.3 and is therefore the characteristic function of an absolutely continuous distribution. Let $p(x)$ be the density of this distribution and write

$$r(x) = \frac{i}{2\pi q} \int_{-\infty}^{\infty} e^{-itx} \psi(t)\, dt.$$

We then have, for $x \neq 0$,

$$p(x) + r(x) = \frac{1}{2\pi} \int_{-\infty}^{\infty} \left[h(t) + \frac{i}{q} \psi(t) \right] e^{-itx}\, dt$$

or

(1.2.6) $\quad p(x) + r(x) = \frac{1}{\pi} \left\{ \int_0^{\infty} h(t) \cos tx\, dt + \frac{1}{q} \int_0^{\infty} \psi(t) \sin tx\, dt \right\}$.

INTRODUCTION

We then have

$$\int_{-\infty}^{\infty} [p(x) + r(x)] e^{itx} \, dx = h(t) + \frac{i}{q} \psi(t),$$

or, on account of (1.2.5a),

$$\int_{-\infty}^{\infty} [p(x) + r(x)] e^{itx} \, dx = \frac{1}{q} [\phi(t) - p + i\psi(t)]$$

so that

(1.2.7) $\quad p + q \int_{-\infty}^{\infty} [p(x) + r(x)] e^{itx} \, dx = \phi(t) + i\psi(t) = f(t).$

It follows from (1.2.4) that

$$\int_{-\infty}^{\infty} [p(x) + r(x)] \, dx = [f(0) - p]/q$$

or, since $\psi(0) = 0$ and $f(0) = \phi(0)$

$$\int_{-\infty}^{\infty} [p(x) + r(x)] \, dx = [\phi(0) - p]/q = [1 - p]/q = 1.$$

In order to show that $f(t)$ is a characteristic function we must prove that $p(x) + r(x)$ is non-negative for all x. Since $p(x)$ is a density, we have $p(0) \geq 0$ and see from (ii) that

$$r(0) = \frac{i}{2\pi q} \int_{-\infty}^{\infty} \psi(t) \, dt = 0.$$

Therefore we have only to show that $p(x) + r(x) \geq 0$ for $x \neq 0$. We introduce the functions

$$I_c(t) = -h(t) - \frac{c}{q} \psi'(t)$$

and

$$J_c(t) = -h(t) - \frac{c}{q} \int_t^{\infty} \psi(t) \, dt.$$

According to assumption (iv) the functions $I_c(t)$ and $J_c(t)$ are concave on $(-\infty, 0)$ and $(0, +\infty)$ for $|c| \leq 1$. Therefore the derivatives of these functions exist almost everywhere and are non-increasing.

We consider the case where $0 < |x| \leq 1$ and integrate the first integral on the right-hand side of (1.2.6) by parts and get

$$\int_0^{\infty} h(t) \cos xt \, dt = -\frac{1}{x} \int_0^{\infty} \sin tx \, h'(t) \, dt.$$

We see then from (1.2.6) that

$$p(x) + r(x) = \frac{1}{\pi x} \int_0^\infty \left[-h'(t) + \frac{x}{q} \psi(t) \right] \sin tx \, dt$$

$$= \frac{1}{\pi x} \int_0^\infty J_x'(t) \sin tx \, dt$$

$$= \frac{1}{\pi x} \sum_{n=0}^\infty \int_{n\pi/x}^{(n+1)\pi/x} J_x'(v) \sin vx \, dv$$

or, putting $v = t + n\pi/x$,

$$p(x) + r(x) = \frac{1}{\pi x} \sum_{n=0}^\infty \int_0^{\pi/x} J_x'(t + n\pi/x) \sin(tx + n\pi) \, dt.$$

Therefore

$$p(x) + r(x) = \frac{1}{\pi x} \sum_{n=0}^\infty \int_0^{\pi/x} \left\{ J_x'\left(t + \frac{2n\pi}{x}\right) - J_x'\left(t + \frac{2n+1}{x}\right) \right\} \sin tx \, dt.$$

Hence

$$p(x) + r(x) \geqslant 0 \quad \text{if} \quad |x| \leqslant 1.$$

We still have to consider the case $|x| \geqslant 1$. In this case we integrate the second integral in (1.2.6) by parts and get

$$\frac{1}{q} \int_0^\infty \psi(t) \sin tx \, dt = \frac{1}{qx} \int_0^\infty \psi'(t) \cos tx \, dt$$

so that

$$p(x) + r(x) = \frac{1}{\pi} \int_0^\infty \left[h(t) + \frac{1}{qx} \psi'(t) \right] \cos tx \, dt$$

$$= \frac{-1}{\pi} \int_0^\infty I_{1/x}(t) \cos tx \, dt.$$

Using partial integration we see that

$$p(x) + r(x) = \frac{1}{\pi x} \int_0^\infty I_{1/x}' \sin tx \, dt$$

and carrying out the same kind of computation as we used in the case $|x| \leqslant 1$ we conclude that

$$p(x) + r(x) = \frac{1}{\pi x} \sum_{n=0}^\infty \int_0^{\pi/x} \left\{ I_{1/x}'\left(t + \frac{2n}{x}\pi\right) - I_{1/x}'\left(t + \frac{2n+1}{\pi}\right) \right\} \sin tx \, dt \geqslant 0.$$

Therefore $p(x) + r(x) \geqslant 0$ and the theorem is proved.

Finally we mention, without proof, a necessary and sufficient condition due to S. M. Berman (1975). This condition assures that a function is the characteristic function of an absolutely continuous distribution.

A complex-valued, measurable function $R(s, t)$ $(-\infty < s, t < \infty)$ is said to be a covariance function if for any positive integer n and any pairs (u_i, u_j) $(i, j = 1, \ldots, n)$

$$\sum_{i=1}^{n} \sum_{j=1}^{n} R(s_i, s_j) u_i u_j \geq 0.$$

For the properties of covariance functions see M. Loève (1977) or Loève's appendix to P. Lévy (1965).

Theorem 1.2.5. Let $R(s, t)$ be a covariance function such that

(1.2.8) $\int R(s, s) \, ds < \infty.$

Then the function

(1.2.9) $r(t) = \dfrac{\displaystyle\int_{-\infty}^{\infty} R(s, s+t) \, ds}{\displaystyle\int_{-\infty}^{\infty} R(s, s) \, ds}$

is a characteristic function.

Berman (1975) also determined the density function corresponding to (1.2.9).

A few somewhat surprising facts are known about the arguments of a characteristic function.

Let $f(t)$ be a characteristic function and let n be a natural number. We say that $f(t)$ has n rolls around the origin if $f(t) \neq 0$ for $t \in [a, b]$ and if

$$\text{arc } f(b) - \text{arc } f(a) \geq n \cdot 2\pi$$

Here arc $f(t)$ is the argument of the characteristic function.

D. O. H. Szász (1973) proved the following theorem.

Theorem 1.2.6. There exists a characteristic function $f(t)$ such that for a suitable T, $f(t) \neq 0$ for $t \in [0, T]$, it is possible to give real numbers $0 \leq a_1 < b_1 \leq a_2 < b_2 \leq \ldots < T$ such that the function $f(t)$ has at least one roll around the origin in each interval $[a_k, b_k]$.

I. V. Ostrovskii & P. M. Flekser (1973) raised the following question: Does there exist a characteristic function $f(t)$ which is different from zero in the open interval (a, b) such that $\int_{-1}^{+1} |\arg f(t)| \, dt = \infty$? They answered it in the affirmative.

1.3 Infinite divisibility

A distribution function $F(x)$ is said to be infinitely divisible if there exists for every positive integer n a distribution function $F_n(x)$ such that $F(x)$ is the n-fold convolution of $F_n(x)$, that is, $F(x) = [F_n(x)]^{*n}$.

In terms of characteristic functions this means that

$$f(t) = [f_n(t)]^n,$$

where $f(t)$ and $f_n(t)$ are the characteristic functions of $F(x)$ and $F_n(x)$ respectively.

Characteristic functions of infinitely divisible distribution functions are called infinitely divisible characteristic functions.

Theorem 1.3.1. *An infinitely divisible characteristic function has no real zeros.*

Theorem 1.3.2. *The product of finitely many infinitely divisible characteristic functions is infinitely divisible.*

The converse statement is not true since there exist infinitely divisible characteristic functions which have factors which are not infinitely divisible.

Theorem 1.3.3. *The limit of a sequence of infinitely divisible characteristic functions is infinitely divisible.*

Infinitely divisible characteristic functions admit canonical representations. We present here a canonical representation due to P. Lévy.

Theorem 1.3.4 (*Lévy canonical representation*). *A necessary and sufficient condition which assures that a characteristic function $f(t)$ is infinitely divisible is that it can be written in the form*

$$\log f(t) = ita - \frac{\sigma^2 t^2}{2} + \int_{-\infty}^{-0} \left(e^{itu} - 1 - \frac{itu}{1+a^2}\right) dM(u)$$

$$+ \int_{+0}^{\infty} \left(e^{itu} - 1 - \frac{itu}{1+u^2}\right) dN(u).$$

Here

(a) $M(u)$ and $N(u)$ are non-decreasing in the intervals $(-\infty, -0)$ and $(0, +\infty)$ respectively;
(b) $M(-\infty) = N(+\infty) = 0$;
(c) the integrals

$$\int_{-\epsilon}^{0} u^2 \, dM(u) \quad \text{and} \quad \int_{0}^{\epsilon} u^2 \, dN(u)$$

are finite for any $\epsilon > 0$, while σ^2 is real and non-negative.

INTRODUCTION

This representation is unique.

Another unique representation of an infinitely divisible characteristic function is the following.

Theorem 1.3.4A (Lévy-Khinchine canonical representation). The function $f(t)$ is an infinitely divisible characteristic function if, and only if, it can be written as

$$\log f(t) = ita + \int_{-\infty}^{\infty} \left(e^{itx} - 1 - \frac{itx}{1+x^2}\right) \frac{1+x^2}{x^2} \, d\theta(x).$$

Here a is a real number while $\theta(x)$ is bounded and non-decreasing, and $\theta(-\infty) = 0$. For $x = 0$ the integrand is defined to be $-t^2/2$.

The function $\theta(x)$ is called the spectral function of the infinitely divisible characteristic function $f(t)$. The functions $M(u)$ and $N(u)$ in Lévy's canonical representation are also called spectral functions. $M(u)$ [resp. $N(u)$] is called the left [resp. right] spectral function.

An important subset of the class of infinitely divisible distributions are the stable distributions. A distribution function $F(x)$ is stable if

$$F\left(\frac{x-a_1}{b_1}\right) * F\left(\frac{x-a_2}{b_2}\right) = F\left(\frac{x-c}{b}\right),$$

where $b_1 > 0$, $b_2 > 0$, a_1, a_2 are real and where c and $b > 0$ are determined by b_1, b_2, a_1, a_2. This equation can be written in terms of characteristic functions as

$$f(b_1 t) f(b_2 t) = f(bt) e^{i\gamma t},$$

where $\gamma = c - c_1 - c_2$. Characteristic functions of stable distributions are called stable characteristic functions.

Theorem 1.3.5. A stable characteristic function admits the canonical representation

(1.3.1) $\log f(t) = iat - c|t|^\alpha \{1 + i\beta \text{ sign } t\omega(|t|, \alpha)\},$

where $c \geq 0$, $|\beta| \leq 1$, $0 < \alpha \leq 2$ and a real. The function

$$\omega(|t|, \alpha) = \begin{cases} \tan\left(\frac{\pi\alpha}{2}\right) & \text{if } \alpha \neq 1 \\ \frac{2}{\pi} \log|t| & \text{if } \alpha = 1. \end{cases}$$

This representation follows from the canonical representation of infinitely divisible characteristic functions. Conversely, a function which admits the representation (1.3.1) is a stable characteristic function.

Analytical properties of stable density functions, in particular their asymptotic expansions and their integral representations, are discussed in detail in Lukacs (1970).

Another important subset of infinitely divisible characteristic functions is the family of self-decomposable characteristic functions.

We defined stable characteristic functions as characteristic functions satisfying the functional equation (1.3.1). In a somewhat similar manner we can define the class of characteristic functions which satisfy the relation

(1.3.2) $\quad f(t) = f(ct) f_c(t)$

for every c ($0 < c < 1$) where $f_c(t)$ is some characteristic function. This class of characteristic functions was introduced by P. Lévy and A. Y. Khinchine. Characteristic functions which belong to this family are called self-decomposable characteristic functions, and the corresponding distribution functions are called self-decomposable distributions. The characteristic functions (distribution functions) of this family are sometimes called the L-class.

Theorem 1.3.6. *All self-decomposable characteristic functions are infinitely divisible.*

Further details about stable and self-decomposable distribution functions (characteristic functions) can be found in Lukacs (1970). A discussion of recent results concerning the L-class is presented in Chapter 5 of Lukacs (1970).

Since the product of two characteristic functions is always a characteristic function, it is clear that some characteristic functions can be written as a product of two (or more) characteristic functions.

A characteristic function $f(t)$ is said to be decomposable if it can be written in the form

$$f(t) = f_1(t) f_2(t),$$

where $f_1(t)$ and $f_2(t)$ are both characteristic functions of non-degenerate distributions. Then $f_1(t)$ and $f_2(t)$ are called factors (or components) of $f(t)$. The restriction to non-degenerate factors is necessitated by the fact that every characteristic function $f(t)$ can be written as a product of two characteristic functions one of which is degenerate. That is, one can always have the trivial decomposition (factorization) $f(t) = f_1(t) f_2(t)$, where $f_1(t) = f(t) e^{iat}$ while $f_2(t) = e^{-iat}$.

Characteristic functions which admit only trivial decompositions are called indecomposable characteristic functions. The terminology introduced here carries over to the corresponding distribution functions.

1.4 Analytic characteristic functions

In this section we consider characteristic functions from the viewpoint of analysis. We note that the characteristic function of the Laplace distribution is

INTRODUCTION

an analytic function, the characteristic function of the normal distribution is an entire function, while the characteristic function of the Cauchy distribution is not analytic.

In view of the strong connection between the values of analytic functions it is convenient to introduce and to study properties of characteristic functions which are analytic.

A characteristic function $f(t)$ is said to be an analytic characteristic function if there exists a function $A(z)$ of the complex variable $z = t + iy$ (t, y real) which is regular in a circle about the origin and for which $A(t) = f(t)$ in some interval containing the point $t = 0$.

Theorem 1.4.1. An analytic characteristic function is always regular in a horizontal strip, called its strip of regularity, and can be represented in this strip by a Fourier-Stieltjes integral.

The strip of regularity can be the whole plane or a half-plane or can have one or two horizontal boundary lines. If the strip has a boundary then the points of intersection of the boundary with the imaginary axis are singular points of the analytic characteristic function. If the strip of regularity is the whole plane then we say that $f(t)$ is an entire characteristic function.

Theorem 1.4.2. Let $f(z)$ be an analytic characteristic function and let iy be a point on the imaginary axis located in the interior of the strip of regularity of $f(z)$. Then

$$|f(t + iy)| \leqslant f(iy)$$

for all real t.

Functions which satisfy this property are called ridge functions. All characteristic functions are ridge functions. However, the converse is not true; there exist ridge functions which are not characteristic functions.

Important results concerning analytic characteristic functions and densities of distributions having analytic characteristic functions can be found in Lukacs (1970).

We call $T(x) = 1 - F(x) + F(-x)$ the tail of the distribution function $F(x)$.

Theorem 1.4.3. Let $F(x)$ be a distribution function and $\alpha > 0$, $k > 0$. Suppose that there exists an $x_0 > 0$ such that the inequality

$$T(x) \leqslant \exp(-kx^{1+\alpha})$$

holds for $x > x_0$. Then $F(x)$ has an entire characteristic function of order equal to or less than $1 + \alpha^{-1}$.

This result is given with proof in Lukacs (1970), Lemma 7.2.2.

We shall also need the following important result due to J. Marcinkiewicz.

Theorem 1.4.4. Let $P_m(t)$ be a polynomial of degree $m > 2$ and denote by $f(t) = \exp[P_m(t)]$. Then $f(t)$ cannot be a characteristic function.

This theorem is given in Lukacs (1970), p. 213; it follows there from a more general result (Theorem 7.3.3).

In view of Theorem 1.4.4 the following result, due to A. A. Goldberg (1973), is of interest.

Theorem 1.4.5. Let $P(t)$ be a preassigned polynomial such that $P(0) = 0$. Then there exist two characteristic functions $f(t)$ and $g(t)$ such that $f(t) = g(t) \exp[P(t)]$ $(-\infty < t < \infty)$.

Theorem 1.4.6. The characteristic function of any non-degenerate, finite distribution function $F(x)$ is an entire function of order 1 and exponential type. The converse statement is also true.

Theorem 1.4.6 is due to G. Pólya (1949) [see also Lukacs (1970), Theorem 7.2.3].

Theorem 1.4.7. Every factor of an entire characteristic function is an entire characteristic function. The order of the factors of an entire characteristic function $f(z)$ cannot exceed the order of $f(z)$ [see Lukacs (1970), Theorem 8.1.2].

We finally give a criterion which assures that a distribution function has an analytic characteristic function.

Theorem 1.4.8. A distribution function $F(x)$ has an analytic characteristic function if, and only if, there exists a positive constant R such that the relation

$$T(x) = 1 - F(x) + F(-x) = O(e^{-rx}) \quad \text{as } x \to \infty$$

holds for all positive $r < R$. The strip of regularity of the characteristic function of $F(x)$ then contains the strip $|\text{Im}(z)| < R$.

2 Moments of arbitrary positive order

In Section 1.1 we discussed moments α_n and β_n whose order was a natural number. It is possible to define moments of a distribution function whose order is not a natural number but an arbitrary positive number. The present chapter deals with properties of such moments.

Let $F(x)$ be a distribution function and let $\lambda > 0$ be a positive real number (not necessarily an integer). We say that $F(x)$ has an absolute moment of order λ if the integral

$$(2.1.1) \quad \beta_\lambda = \int_{-\infty}^{\infty} |x|^\lambda \, dF(x)$$

exists and is finite. In this case the (algebraic) moment

$$(2.1.2) \quad \alpha_\lambda = \int_{-\infty}^{\infty} x^\lambda \, dF(x)$$

exists also.

2.1 Existence of moments; behaviour of distribution functions and characteristic functions

In this section we show that the existence of moments of a distribution function is closely related to the behaviour of its characteristic function in the neighbourhood of the origin and to the behaviour of the distribution function for arguments of large absolute value.

Theorem 2.1.1. *Let $F(x)$ be a distribution function with characteristic function $f(t)$. Suppose that there exists a sequence $\{t_n\}$ and a constant λ, $0 < \lambda < 2$, such that*

(i) $\lim_{n \to \infty} t_n = 0$

(ii) $\sum_n |t_n|^\epsilon$ *converges for any $\epsilon > 0$*

(iii) *the sequence $\{t_{n-1}/t_n\}$ is bounded*

(iv) $\log |f(t_n)|/|t_n|^\lambda$ *is bounded.*

Then $F(x)$ has absolute moments of any order inferior to λ.

Remark. The conditions of the theorem can be satisfied if $\log|f(t)|/|t|^\lambda$ is bounded in $0<|t|<\eta$ for some η (i.e. in a deleted neighbourhood of 0). This is immediately seen if one takes $t_n = \theta^n$, where $0<\theta<1$.

Without loss of generality we can assume that $1>t_{n-1}>t_n>0$ for all $n \geqslant 2$. According to our assumptions, there exists a positive constant c_1 such that
$$1-|f(t_n)|^2 \leqslant 1 - \exp(-c_1 t_n^\lambda) \leqslant c_1 t_n^\lambda$$
for n sufficiently large. We saw in Theorem 1.1.5 that $|f(t)|^2 = f(t)f(-t)$ is the characteristic function of the convolution of $F(x)$ with its conjugate distribution $\tilde{F}(x)$. We write $F^* = F * \tilde{F}$ and see that
$$|f(t_n)|^2 = \int_{-\infty}^{\infty} \cos t_n x \, dF^*(x).$$

We put $u_n = t_n/2$, so that
$$\int_{-\infty}^{\infty} \sin^2 u_n x \, dF^*(x) = \tfrac{1}{2}(1-|f(2u_n)|^2) \leqslant c_2 u_n^\lambda.$$

We note that $\sin^2 \theta \geqslant \theta^2 \sin^2 1$ for $0 \leqslant \theta \leqslant 1$ and putting $x_n = 1/u_n$ we see that
$$\int_{x_{n-1}}^{x_n} x^2 \, dF^*(x) \leqslant c_3 \int_{x_{n-1}}^{x_n} \frac{\sin^2(u_n x)}{u_n^2} \, dF^*(x) \leqslant c_4 u_n^{\lambda-2}.$$

Let $0<\delta<\lambda$; for $x \in [x_{n-1}, x_n]$ one has $x^\delta \leqslant x_n^\delta x^2/x_{n-1}^2$, so that
$$\int_{x_{n-1}}^{x_n} x^\delta \, dF^*(x) \leqslant \left(\frac{x_n^\delta}{x_{n-1}^2}\right) \int_{x_{n-1}}^{x_n} x^2 \, dF^*(x) \leqslant c_4 \frac{x_n^\delta}{x_{n-1}^2} u_n^{\lambda-2}$$
or
$$\int_{x_{n-1}}^{x_n} x^\delta \, dF^*(x) \leqslant c_4 \left(\frac{u_{n-1}}{u_n}\right)^2 u_n^{\lambda-\delta}.$$

It follows from assumption (iii) that
$$\int_{x_{n-1}}^{x_n} x^\delta \, dF^*(x) \leqslant c_5 u_n^{\lambda-\delta}.$$

Finally we conclude from assumption (ii) that
$$\int_{-\infty}^{\infty} x^\delta \, dF^*(x) < \infty$$
so that F^*, and therefore also F, have moments of all orders inferior to λ.

MOMENTS OF ARBITRARY POSITIVE ORDER

Remark. Condition (iv) is equivalent to the statement that

(iv*) $\quad \dfrac{1-|f(t_n)|^2}{|t_n|^\lambda}$ is bounded

and also to the statement that

(iv**) $\quad \dfrac{1-\operatorname{Re} f(t_n)}{|t_n|^\lambda}$ is bounded.

To prove the equivalence of (iv*) and (iv**) we note that

(#) $\quad \dfrac{1-|f(t)|^2}{|t|^\lambda} = \dfrac{1-\operatorname{Re} f(t)}{|t|^\lambda}[1+\operatorname{Re} f(t)] - \left(\dfrac{\operatorname{Im} f(t)}{|t|^{\lambda/2}}\right)^2.$

It follows from relation (#) that (iv**) implies (iv*). On the other hand, if (iv*) holds one can conclude from Theorem 2.1.1 that β_k exists for $k < \lambda$. Hence

$$\dfrac{|\operatorname{Im} f(t)|}{|t|^{\lambda/2}} = \left|\int \dfrac{\sin tx}{|t|^{\lambda/2}} \, dF(x)\right| \leq \int \dfrac{|\sin tx|}{|tx|^{\lambda/2}} |x|^{\lambda/2} \, dF(x)$$

$$\leq \int |x|^{\lambda/2} \, dF(x) = \beta_{\lambda/2}.$$

The boundedness of $\dfrac{|\operatorname{Im} f(t)|}{|t|^{\lambda/2}}$ together with the boundedness of $\dfrac{1-|f(t)|^2}{|t|^\lambda}$ mean, on account of (#) that $\dfrac{1-\operatorname{Re} f(t)}{|t|^\lambda}$ is also bounded.

The theorem is due to B. Ramachandran & C. R. Rao (1968). Related results can be found in Ramachandran (1969).

The next theorem gives the connection between the existence of moments of a distribution function $F(x)$ and its "tail behaviour", that is, the behaviour of $1 - F(x)$ and $F(-x)$ for values of the argument whose absolute value $|x|$ is large.

Theorem 2.1.2. Suppose that the absolute moment β_λ ($\lambda > 0$) of a distribution function $F(x)$ exists. Then

(2.1.3) $\quad 1 - F(x) + F(-x) = o(x^{-\lambda}) \quad (x \to \infty).$

Conversely, if relation (2.1.3) holds then β_μ exists for all $\mu < \lambda$.

Remark. The validity of the relation (2.1.3) does not permit the conclusion that β_λ exists. This is illustrated by the following example.

Let
$$p(x) = \begin{cases} 0 & \text{for } x < 2 \\ \dfrac{c}{x^{\lambda+1} \log x} & \text{for } x \geq 2 \end{cases}$$

where $c^{-1} = \int_2^\infty \dfrac{dx}{x^{\lambda+1} \log x}$. Then

$$1 - F(x) + F(-x) = 1 - F(x) = \int_x^\infty p(y)\, dy = \int_x^\infty \dfrac{c}{y^{\lambda+1} \log y}\, dy$$

$$< \dfrac{c}{\log x} \int_x^\infty y^{-\lambda-1}\, dy = \dfrac{c}{\lambda} \dfrac{x^{-\lambda}}{\log x} = o(x^{-\lambda}) \quad \text{as } x \to \infty,$$

while the integral $\int_2^\infty y^\lambda p(y)\, dy = c \int_2^\infty \dfrac{dy}{y \log y}$ does not exist (finitely).

To prove the theorem we assume first that β_λ exists. Then

$$\int_R^\infty x^\lambda\, dF(x) = o(1) \quad \text{and} \quad \int_{-\infty}^{-R} |x|^\lambda\, dF(x) = o(1) \text{ as } R \to \infty$$

(2.1.4a) $\quad R^\lambda \int_R^\infty dF(x) \leq \int_R^\infty x^\lambda\, dF(x) = o(1) \quad$ or

$$\int_R^\infty dF(x) = 1 - F(R) = o(R^{-\lambda}).$$

Similarly,

$$R^\lambda F(-R) = R^\lambda \int_{-\infty}^{-R} dF(x) \leq \int_{-\infty}^{-R} |x|^\lambda\, dF(x) = o(1), \text{ so that}$$

(2.1.4b) $\quad F(-R) = o(R^{-\lambda})$.

Formula (2.1.3) follows from (2.1.4a) and (2.1.4b) so that the first statement of the theorem is proved.

We now assume that (2.1.3) holds, then

(2.1.5a) $\quad \lim_{R \to \infty} R^\lambda H(R) = 0$

where $H(x) = 1 - F(x)$ and

(2.1.5b) $\quad \lim_{R \to \infty} R^\lambda F(-R) = 0$.

It follows from (2.1.5a) that for any $\epsilon > 0$ and R, $R^\lambda H(R) < \epsilon$ for R sufficiently large, say $R > R_1$. Let $0 < \mu < \lambda$, then

$$\int_R^\infty x^\mu \, dF(x) = -\int_R^\infty x^\mu \, dH(x) = R^\mu H(R) + \mu \int_R^\infty x^{\mu-1} H(x) \, dx.$$

One has $R^\mu H(R) < R^\lambda H(R) < \epsilon$ and also

$$\int_R^\infty x^{\mu-1} H(x) \, dx < \epsilon \int_R^\infty x^{-(\lambda-\mu)-1} \, dx = \epsilon \frac{R^{-(\lambda-\mu)}}{\lambda - \mu}.$$

Therefore

(2.1.6a) $\quad \lim\limits_{R \to \infty} \int_R^\infty x^\mu \, dF(x) = \lim\limits_{R \to \infty} \int_R^\infty |x|^\mu \, dF(x) = 0.$

In a similar way one can show that (2.1.5b) implies that

$$\int_{-\infty}^{-R} |x|^\mu \, dF(x) < \epsilon + \frac{\epsilon \mu}{\lambda - \mu} R^{-(\lambda - \mu)}$$

so that

(2.1.6b) $\quad \lim\limits_{R \to \infty} \int_{-\infty}^{-R} |x|^\mu \, dF(x) = 0.$

The existence of β_μ with $\mu < \lambda$ follows immediately from (2.1.6a) and (2.1.6b).

2.2 Expansion of characteristic functions

We first derive an expansion for the characteristic function of a distribution function which has moments of order $\lambda = n + \delta$, where n is a non-negative integer while $0 < \delta \leq 1$. This expansion is analogous to the expansion given by formula (2.3.4a) of Lukacs (1970) from which it differs only in the remainder term.

Let $F(x)$ be a distribution function with characteristic function $f(t)$. If the moment $\beta_{n+\delta}$ ($0 < \delta \leq 1$) exists, then the moments of all orders $1, 2, \ldots, n$ exist and $f(t)$ can be differentiated n times. For every positive integer $k \leq n$,

$$f^{(k)}(t) = i^k \int_{-\infty}^\infty e^{itx} x^k \, dF(x)$$

and one has the Maclaurin expansion

(2.2.1) $\quad f(t) = \sum\limits_{k=0}^{n-1} \alpha_k \frac{(it)^k}{k!} + R_n.$

The remainder term is given by the formula[*]

$$R_n = R_n(t) = \frac{t^n}{(n-1)!} \int_0^1 (1-u)^{n-1} f^{(n)}(tu) \, du$$

We compute the integral in the remainder term and note first that

$$f^{(n)}(tu) = i^n \int_{-\infty}^{\infty} (e^{iutx} - 1) x^n \, dF(x) + i^n \alpha_n,$$

so that

(2.2.2)
$$R_n(t) = \frac{t^n}{(n-1)!} \int_0^1 (1-u)^{n-1} \int_{-\infty}^{\infty} (e^{iutx} - 1) x^n \, dF(x) \, du + \frac{(it)^n}{n!} \alpha_n.$$

For the estimation of the integral $\int_{-\infty}^{\infty} (e^{iutx} - 1) x^n \, dF(x)$ we need the following lemma.

Lemma 2.2.1. *If $0 < \delta \leq 1$ then $|e^{ia} - 1| \leq 2|a/2|^\delta$.*

For the proof of the lemma we distinguish two cases: If $|a/2| < 1$ we see easily that $|e^{ia} - 1| = 2|\sin(a/2)| \leq 2|a/2| \leq 2|a/2|^\delta$, while for $|a/2| \geq 1$ one has $|e^{ia} - 1| \leq 2 \leq 2|a/2|^\delta$.

Using this estimate we see that

$$\left| \int_{-\infty}^{\infty} (e^{iutx} - 1) x^n \, dF(x) \right| \leq 2^{1-\delta} u^\delta |t|^\delta \beta_{n+\delta}.$$

We substitute this into (2.2.2) and get

$$|R_n(t)| \leq 2^{1-\delta} |t|^{n+\delta} \beta_{n+\delta} \int_0^1 \frac{(1-u)^{n-1}}{(n-1)!} u^\delta \, du + \frac{(it)^n}{n!} \alpha_n.$$

We integrate repeatedly by parts and see finally that

$$|R_n(t)| \leq \frac{2^{1-\delta} \beta_{n+\delta} |t|^{n+\delta}}{(1+\delta)(2+\delta)\ldots(n+\delta)} + \frac{(it)^n}{n!} \alpha_n.$$

Hence

(2.2.3) $\quad f(t) = \sum_{k=0}^{n} \alpha_k \frac{(it)^k}{k!} + \theta \, \frac{2^{1-\delta} \beta_{n+\delta} |t|^{n+\delta}}{(1+\delta)(2+\delta)\ldots(n+\delta)}$

where $|\theta| \leq 1$.

---

[*] See again Hardy (1963).

MOMENTS OF ARBITRARY POSITIVE ORDER

We have therefore obtained the following result:

Theorem 2.2.1. Let $F(x)$ be a distribution function and suppose that its moment $\beta_{n+\delta}$ (n non-negative integer, $0 < \delta \leqslant 1$) exists. Then the characteristic function $f(t)$ of $F(x)$ admits an expansion of the form

$$(2.2.4) \quad f(t) = \sum_{k=0}^{n} \alpha_k \frac{(it)^k}{k!} + O(|t|^{n+\delta})$$

as $|t| \to 0$ and where $0 < \delta \leqslant 1$.

We next show that a distribution function whose characteristic function admits an expansion of the form

$$(2.2.4a) \quad f(t) = \sum_{k=0}^{n} a_k \frac{(it)^k}{k!} + O(|t|^{n+\delta})$$

has moments of all orders inferior to $n + \delta$.

Let $F(x)$ be a distribution function and let $F^* = F * \tilde{F}$ be the corresponding symmetrized distribution. The characteristic function $f^*(t)$ of $F^*(x)$ is then real and is given by $f^*(t) = f(t)f(-t) = |f(t)|^2$. We know that F has moments of order λ if F^* has moments of order λ. Without loss of generality we can therefore consider the existence of moments of the symmetrized distribution function F^*. Since $F^*(x)$ is a symmetric distribution, all moments of F^* whose order is an odd integer are zero and the expansion (2.2.4a) can be written in the form

$$(2.2.5) \quad f^*(t) = 1 + \sum_{j=1}^{m} a_{2j} \frac{(it)^{2j}}{(2j)!} + O(|t|^{2m+\delta}) \quad (t \to 0)$$

where

$$(2.2.5a) \quad 0 < \delta \leqslant 2.$$

If $m = 0$, then it follows from Theorem 2.1.1 that $F^*(x)$ has moments of all orders inferior to δ. Therefore we assume from now on that $m > 0$.

Let Δ_n^h be the symmetric difference operator defined by

$$\Delta_n^h g(y) = \sum_{k=0}^{n} (-1)^k \binom{n}{k} g[y + (n - 2k)h].$$

Then

$$(2.2.6) \quad \Delta_n^h e^{ixt} = e^{ixt}[2i \sin xh]^n.$$

We apply the operator Δ_{2m}^h to the left-hand side of (2.2.5), noting that $f^*(t)$ is a real-valued function, and get

$$(2.2.6a) \quad \Delta_{2m}^h f^*(t) = (-1)^m 2^{2m} \int_{-\infty}^{\infty} \cos tx [\sin hx]^{2m} \, dF^*(x).$$

Before applying the operator Δ_{2m}^h to the right-hand side of (2.2.5) we note that

(2.2.7) $\quad \Delta_n^h t^n = 2^n n! \, h^n$

while for $k < n$

$$\Delta_n^h t^k = 0.$$

If we apply Δ_{2m}^h to the right-hand side of (2.2.5) we get

(2.2.8) $\quad \Delta_{2m}^h f^*(t) = \dfrac{(-1)^m}{(2m)!} \alpha_{2m} \Delta_{2m}^h t^{2m} + A(h, t)$

$$= (-1)^m 2^{2m} \alpha_{2m} h^{2m} + A(h, t)$$

where

(2.2.8a) $\quad A(h, h) = O(|h|^{2m+\delta}) \quad (|h| \to 0)$.

We equate (2.2.6) and (2.2.8) and see after an elementary computation that

$$\int_{-\infty}^{\infty} \cos hx \, [\sin hx]^{2m} \, dF^*(x) = \alpha_{2m} h^{2m} + O(|h|^{2m+\delta}).$$

Therefore

$$\int_{-\infty}^{\infty} \cos hx \left[\frac{\sin hx}{hx}\right]^{2m} x^{2m} \, dF^*(x) = \alpha_{2m} + O(|h|^\delta)$$

so that

$$\int_{-\infty}^{\infty} \left\{1 - \cos hx \left[\frac{\sin hx}{hx}\right]^{2m}\right\} x^{2m} \, dF^*(x) = O(|h|^\delta).$$

Hence

(2.2.9) $\quad \displaystyle\int_{-\infty}^{\infty} \left\{1 - \left[\frac{\sin hx}{hx}\right]^{2m}\right\} x^{2m} \, dF^*(x) = O(|h|^\delta)$

as $h \to 0$.

Let $F_{2m}(x) = \dfrac{1}{\alpha_{2m}} \displaystyle\int_{-\infty}^{x} y^{2m} \, dF^*(y)$; clearly $F_{2m}(x)$ is a distribution function and it follows from (2.2.9) that

$$\int_{-\infty}^{\infty} \left\{1 - \left[\frac{\sin hx}{hx}\right]^{2m}\right\} dF_{2m}(x) = O(|h|^\delta) \quad \delta \to 0.$$

Since $\sin hx / hx \leq \tfrac{1}{2}$ for $hx \geq 2$, and since

$$\int_{-\infty}^{\infty} \left\{1 - \left[\frac{\sin hx}{hx}\right]^{2m}\right\} dF_{2m}(x) \geq \int_{hx \geq 2} \left\{1 - \left[\frac{\sin hx}{hx}\right]^{2m}\right\} dF_{2m}(x)$$

we conclude that

$$(1 - 2^{-2m})\left[1 - F_{2m}\left(\frac{2}{h}\right)\right] = O(|h|^\delta) \quad \text{as } \delta \to 0$$

or

(2.2.10a) $\quad 1 - F_{2m}(y) = O(y^{-\delta}) \quad \text{as } y \to \infty.$

In a similar way one can show that

(2.2.10b) $\quad F_{2m}(-y) = O(y^{-\delta}) \quad \text{as } y \to \infty.$

It follows then from Theorem 2.1.2 that $F_{2m}(x)$ has all moments of order inferior to δ. Then $F^*(x)$, and therefore also $F(x)$, have moments of all orders inferior to $2m + \delta$. Thus we have obtained the following result:

Theorem 2.2.2. Let $F(x)$ be a distribution function whose characteristic function $f(t)$ admits an expansion of the form

$$f(t) = \sum_{k=0}^{n} a_k \frac{(it)^k}{k!} + O(|t|^{n+\delta}) \quad (0 < \delta \leq 1)$$

as $t \to 0$. Then $F(x)$ has all moments of orders inferior to $n + \delta$.

Theorems 2.2.1, 2.2.2 and other related results can be found in Ramachandran (1969).

2.3 Fractional moments and fractional derivatives

It follows from Theorem 2.2.1 that moments and absolute moments of a distribution function can be expressed in terms of derivatives of the corresponding characteristic function. In this section we derive a similar result for moments whose order is positive but not a non-negative integer. To accomplish this aim one has to introduce a new concept, namely derivatives of non-integer orders. One refers often, in a somewhat sloppy way, to moments and derivatives of positive non-integer order as "fractional moments" and "fractional derivatives", although the order need not be a rational number. Work similar to fractional differentiation exists also in connection with integration, and one refers to these studies collectively as fractional calculus.

An extensive study of fractional calculus was carried out by A. Marchaud (1927).

Let k be a non-negative integer and $0 < \lambda < 1$. The fractional derivative of order $k + \lambda$ of a function $f(t)$ is defined in the following way:

(2.3.1) $\quad \dfrac{d^{k+\lambda}}{dt^{k+\lambda}} f(t) = \dfrac{\lambda}{\Gamma(1-\lambda)} \int_{-\infty}^{t} \dfrac{f^{(k)}(t) - f^{(k)}(u)}{(t-u)^{1+\lambda}} \, du.$

Here $f^{(k)}(t)$ is the kth derivative of $f(t)$. To simplify the notation we shall often write $D^{k+\lambda}$ instead of $\dfrac{d^{k+\lambda}}{dt^{k+\lambda}}$.

It follows from (2.3.1) that

$$D^{k+\lambda} f(t)|_{t=0} = \frac{(-1)}{\Gamma(-\lambda)} \int_{-\infty}^{0} \frac{f^{(k)}(0) - f^{(k)}(u)}{(-u)^{1+\lambda}} du$$

$$= \frac{(-1)}{\Gamma(-\lambda)} \int_{0}^{\infty} \frac{f^{(k)}(0) - f^{(k)}(-u)}{u^{1+\lambda}} du.$$

If $f(t)$ is the characteristic function of the distribution $F(x)$ then we have

$$(2.3.2) \quad (-i)^k D^{k+\lambda} f(t)|_{t=0} = \frac{(-1)}{\Gamma(-\lambda)} \left\{ \int_0^\infty \left[\int_{-\infty}^\infty x^k (1 - \cos ux)\, dF(x) \right. \right.$$

$$\left. \left. + i \int_{-\infty}^\infty x^k \sin ux\, dF(x) \right] \frac{du}{u^{1+\lambda}} \right\}.$$

Formula (2.3.2) is valid for even as well as for odd integers k.

The expressions which we shall obtain for fractional moments, as well as the derivatives, are different for even and for odd k. Therefore we formulate the results as two theorems.

Theorem 2.3.1. Let $F(x)$ be a distribution function with characteristic function $f(t)$, and let $k = 2n$ be an even integer and $0 < \lambda < 1$. The absolute moment $\beta_{k+\lambda} = \beta_{2n+\lambda}$ of $F(x)$ exists if, and only if,

(i) $\beta_{2n} < \infty$

(ii) $\mathrm{Re}\left[\dfrac{d^{2n+\lambda}}{dt^{2n+\lambda}} f(t)\bigg|_{t=0} \right]$ exists;

then

$$(2.3.3) \quad \beta_{2n+\lambda} = \frac{1}{\cos \lambda \pi/2} \mathrm{Re}\left[(-1)^n \frac{d^{2n+\lambda}}{dt^{2n+\lambda}} f(t)\bigg|_{t=0} \right].$$

Here $\mathrm{Re}\,[\ ]$ is the real part of the expression in the brackets.

For the proof of the theorem we need the following lemma.

Lemma 2.3.1

(a) $\displaystyle\int_0^\infty \frac{1 - \cos v}{v^{1+\beta}} dv = -\Gamma(-\beta) \cos \frac{\pi \beta}{2} \quad (0 < \beta < 2)$

(b) $\displaystyle\int_0^\infty \frac{1 - \cos v}{v^{2+\alpha}} dv = \frac{\Gamma(-\alpha)}{1+\alpha} \sin \frac{\pi \alpha}{2} \quad (|\alpha| < 1).$

We prove the lemma by using a known formula [see Kawata (1972), p. 430]:

$$\int_0^\infty \frac{1-\cos xt}{t^{1+\beta}}\,dt = \frac{\Gamma(2-\beta)}{\beta(1-\beta)}\sin\left[(1-\beta)\frac{\pi}{2}\right]|x|^\beta \qquad 0<\beta<2.$$

We note that $\Gamma(2-\beta) = -\beta(1-\beta)\Gamma(-\beta)$ and transposing $|x|^\beta$ to the left-hand side of the formula we obtain (a) by introducing the new variable $v = |x|t$. Statement (b) then follows easily by putting $\beta = 1 + \alpha$.

We proceed to prove Theorem 2.3.1 and putting $k = 2n$ we obtain from (2.3.2)

(2.3.4) $\operatorname{Re}[(-1)^n D^{2n+\lambda} f(t)|_{t=0}]$

$$= \frac{-1}{\Gamma(-\lambda)} \int_0^\infty \left[\int_{-\infty}^\infty |x|^{2n}(1-\cos ux)\,dF(x)\right] \frac{du}{u^{1+\lambda}}$$

$$= -\frac{1}{\Gamma(-\lambda)} \int_{-\infty}^\infty |x|^{2n+\lambda} \left[\int_0^\infty \frac{1-\cos u|x|}{(u|x|)^{1+\lambda}} |x|\,du\right] dF(x).$$

We introduce $v = u|x|$ in the inner integral and use lemma 2.3.1 and (2.3.4) and see that conditions (i) and (ii) of the theorem are sufficient for the validity of (2.3.3). To prove the necessity of conditions (i) and (ii) one has only to note that the reasoning presented above can be reversed.

Before discussing the case where k is an odd integer we must make the following remark.

Remark. Let $F(x)$ be a distribution function whose absolute moment β_{2n} of order $2n$ exists. Then

$$H_{2n}(x) = \frac{1}{\beta_{2n}} \int_{-\infty}^x y^{2n}\,dF(y)$$

is also a distribution function. The characteristic function of $H_{2n}(x)$ is

(2.3.5) $h_{2n}(t) = \dfrac{1}{\beta_{2n}} \displaystyle\int_{-\infty}^\infty e^{itx} x^{2n}\,dF(x) = \dfrac{1}{(-1)^n \beta_{2n}} f^{(2n)}(t),$

where $f^{(2n)}(t)$ is the $2n$th derivative of the characteristic function $f(t)$ of $F(x)$.

Theorem 2.3.2. *Let $F(x)$ be a distribution function with characteristic function $f(t)$, and let $k = 2n + 1$ be an odd integer and $0 < \lambda < 1$. The absolute moment $\beta_{2n+1+\lambda}$ of $F(x)$ exists if and only if*

(i) $\beta_{2n+1} < \infty$;
(ii) $\operatorname{Re}[D^{2n+1+\lambda} f(t)|_{t=0}]$ *exists*;
(iii) $\lim\limits_{t \to 0} \dfrac{1 - \operatorname{Re} h_{2n}(t)}{t^{1+\lambda}}$ *exists*.

Then

(2.3.6) $\quad \beta_{2n+1+\lambda} = \dfrac{1}{\sin(\pi\lambda/2)} \operatorname{Re}[(-1)^{n+1} D^{2n+1+\lambda} f(t)|_{t=0}].$

Here $h_{2n}(t)$ is given by (2.3.5).

For the proof of the theorem we need the following lemma:

Lemma 2.3.2. *Let $F(x)$ be a distribution function with characteristic function $f(t)$, let k be a positive integer, and let λ be a real number such that $0 < \lambda < 1$. The function $f(t)$ admits an expansion*

$$f(t) = 1 + \sum_{j=1}^{k} \alpha_j (it)^j / j! + o(|t|^{k+\lambda}) \quad (\text{as } t \to 0)$$

if and only if

$$1 - F(x) + F(-x) = o(x^{-(k+\lambda)}) \quad (\text{as } x \to \infty).$$

For the proof we refer to S. J. Wolfe (1973). From (2.3.2) we see that

(2.3.7) $\quad \operatorname{Re}[D^{2n+1+\lambda} f(t)|_{t=0}] = \dfrac{(-1)^n}{\Gamma(-\lambda)} \displaystyle\int_0^\infty \left[\int_{-\infty}^\infty x^{2n+1} \sin ux \, dF(x) \right] \dfrac{du}{u^{1+\lambda}}.$

We now introduce the characteristic function $h_{2n}(t)$, given by (2.3.5). According to assumption (i) of the theorem, β_{2n+1}, and therefore also $f^{(2n+1)}(t)$, exist. It follows from (2.3.5) that

$$\dfrac{d}{dt}[\operatorname{Re} h_{2n}(t) - 1] = \dfrac{(-1)^n}{\beta_{2n}} \operatorname{Re} f^{(2n+1)}(t) = \dfrac{-1}{\beta_{2n}} \int_{-\infty}^\infty x^{2n+1} \sin xt \, dF(x).$$

We substitute the last expression into (2.3.7) and get

(2.3.8) $\quad \operatorname{Re}[(-1)^{n+1} D^{2n+1+\lambda} f(t)|_{t=0}]$

$$= \dfrac{\beta_{2n}}{\Gamma(-\lambda)} \int_0^\infty \dfrac{(d/du)[\operatorname{Re} h_{2n}(u) - 1]}{u^{1+\lambda}} \, du.$$

Integration by parts yields

(2.3.9) $\quad \displaystyle\int_0^\infty \dfrac{(d/du)[\operatorname{Re} h_{2n}(u) - 1]}{u^{1+\lambda}} \, du$

$$= \left[\dfrac{\operatorname{Re} h_{2n}(u) - 1}{u^{1+\lambda}} \right]_0^\infty + \Gamma(1+\lambda) \int_0^\infty \dfrac{[1 - \operatorname{Re} h_{2n}(u)]}{u^{2+\lambda}} \, du.$$

MOMENTS OF ARBITRARY POSITIVE ORDER

In view of assumptions (ii) and (iii) we conclude that the integral

$$(2.3.10) \quad \int_0^\infty \frac{[1 - \operatorname{Re} h_{2n}(u)]}{u^{2+\lambda}} \, du$$

exists. It follows from (2.3.5) that

$$(2.3.11) \quad 1 - \operatorname{Re} h_{2n}(u) = \frac{1}{\beta_{2n}} \int_{-\infty}^\infty x^{2n}(1 - \cos ux) \, dF(x).$$

We combine (2.3.10) and (2.3.11) and see that

$$J = \int_0^\infty \frac{1 - \operatorname{Re} h_{2n}(u)}{u^{2+\lambda}} \, du = \frac{1}{\beta_{2n}} \int_0^\infty \left[\int_{-\infty}^\infty x^{2n}(1 - \cos ux) \, dF(x) \right] \frac{du}{u^{2+\lambda}}.$$

We exchange the order of integration and write $|x|^{2n}$ instead of x^{2n} and get

$$J = \frac{1}{\beta_{2n}} \int_{-\infty}^\infty |x|^{2n+1+\lambda} \, dF(x) \int_0^\infty \frac{1 - \cos u|x|}{(u|x|)^{2+\lambda}} \, |x| \, du.$$

We introduce the new variable $v = u|x|$ and using formula (b) of Lemma 2.3.1 we obtain

$$(2.3.12) \quad \int_0^\infty \frac{1 - \operatorname{Re} h_{2n}(u)}{u^{2+\lambda}} \, du = \frac{1}{\beta_{2n}} \beta_{2n+1+\lambda} \frac{\Gamma(-\lambda)}{1+\lambda} \sin \frac{\pi\lambda}{2}.$$

We already know that the integral on the left-hand side of (2.3.12) exists and conclude therefore that the moment $\beta_{2n+1+\lambda}$ also exists.

In order to derive the expression (2.3.6) for $\beta_{2n+1+\lambda}$ we have to combine (2.3.8) and (2.3.9), so that

$$\operatorname{Re}[(-1)^{n+1} D^{2n+1+\lambda} f(t)|_{t=0}]$$
$$= \frac{\beta_{2n}}{\Gamma(-\lambda)} \left[\left\{ \frac{\operatorname{Re} h_{2n}(u) - 1}{u^{1+\lambda}} \right\}_0^\infty + (1+\lambda) \int_0^\infty \frac{1 - \operatorname{Re} h_{2n}(u)}{u^{2+\lambda}} \, du \right].$$

We combine this with (2.3.12) and get

$$\operatorname{Re}[(-1)^{n+1} D^{2n+1+\lambda} f(t)|_{t=0}] =$$
$$\frac{\beta_{2n}}{\Gamma(-\lambda)} \left[\frac{\operatorname{Re} h_{2n}(u) - 1}{u^{1+\lambda}} \right]_0^\infty + \beta_{2n+1+\lambda} \sin \frac{\pi\lambda}{2}.$$

In order to obtain (2.3.6) we must show that

$$(2.3.13) \quad \left[\frac{\operatorname{Re} h_{2n}(u) - 1}{u^{1+\lambda}} \right]_0^\infty = \lim_{u \to 0} \frac{\operatorname{Re} h_{2n}(u) - 1}{u^{1+\lambda}} = 0.$$

If the moment $\beta_{2n+1+\lambda}$ of a distribution function $F(x)$ exists, then, according to Theorem 2.1.2, $1-F(x)+F(-x)=o(x^{-\lambda})$ as $x \to \infty$. One can therefore apply Lemma 2.3.2 and conclude that $f(t)$ admits the expansion

$$f(t) = 1 + \sum_{j=1}^{2n+1} \alpha_j (it)^j / j! + o(|t|^{2n+1+\lambda})$$

as $t \to 0$. Hence

$$\lim_{t \to 0} \frac{\mathrm{Re}\left[f(t) - \sum_{j=1}^{2n+1} \alpha_j (it)^j / j! \right]}{t^{2n+1+\lambda}} = 0.$$

We apply de l'Hôpital's rule $2n$ times and note that $f^{(2n)}(t) = (-1)^n \alpha_{2n} h_{2n}(t)$; then

$$\lim_{t \to 0} \frac{\mathrm{Re}\,(-1)^n \alpha_{2n} h_{2n}(t) - (-1)^n \alpha_{2n}}{C_1 t^{1+\lambda}} = 0,$$

that is, $\lim\limits_{t \to 0} C\, \dfrac{\mathrm{Re}\, h_{2n}(t) - 1}{t^{1+\lambda}} = 0$ (C_1, C constants) and (2.3.13) follows immediately.

To prove the necessity of the conditions, we note that the assumption that $\beta_{2n+1+\lambda}$ exists implies the existence of β_{2n+1} (that is, condition (i)). We also saw that condition (iii) follows from the existence of $\beta_{2n+1+\lambda}$, while (ii) follows from the representation (2.3.6).

The argument used in the proofs of Theorems 2.3.1 and 2.3.2 is due to G. Laue (1980).

S. J. Wolfe (1975) carried out a similar study but also considered moments of negative order by using fractional integrals.

3 Estimation of the closeness of distribution functions

In view of the continuity theorem one can expect that two distribution functions whose characteristic functions do not differ much are also close to each other in some sense. To obtain precise statements it is necessary to introduce a method of measuring the closeness of distribution functions. It appears that it is possible to consider the set of all distribution functions as a metric space.(*) There exist many ways in which a metric can be defined in the set of distribution functions. In this book we shall deal with two distance definitions. Systematic studies of metrics in the space of distribution functions have been made by several authors. We mention here only V. V. Senatov (1977) and V. M. Zolotarev (1976).

3.1 Distance definitions

Next we introduce the two distance definitions which we wish to consider and briefly discuss some of their properties.

Let $F(x)$ and $G(x)$ be two distribution functions, then it is easily seen that

(3.1.1) $\rho(F, G) = \sup_x |F(x) - G(x)|$

is a metric so that the set of all distribution functions becomes a metric space. The metric (3.1.1) is called the uniform metric; alternatively the term Kolmogorov metric is also used.

The second metric which we shall use was proposed by P. Lévy (1937a).

Let $F(x)$ and $G(x)$ be two distribution functions; their distance $L(F, G)$ is defined as the infimum of all positive h which satisfy, for all x, the inequality

(3.1.2) $F(x - h) - h \leq G(x) \leq F(x + h) + h$.

It is not difficult to prove that $L(F, G)$ is a metric in the space of distribution functions. For the proof we refer to B. V. Gnedenko & A. N. Kolmogorov (1954). It is customary to call $L(F, G)$ the Lévy distance.

The Lévy metric is a very natural tool in probability theory since it has the following property:

(*) A set X is said to be a metric space if for every pair of points $x, y \in X$ a distance (metric) $d(x, y)$ is defined which has the following properties:

    (i) $d(x, y) = d(y, x)$ (symmetry property).
    (ii) $d(x, y) \geq 0$ and $d(x, y) = 0$ if and only if $x = y$ (identification property).
    (iii) $d(x, y) + d(y, z) \geq d(x, z)$ (triangle inequality).

Theorem 3.1.1. *Let $\{F_n(x)\}$ be a sequence of distribution functions. This sequence converges weakly*[*] *to a distribution function $F(x)$ if and only if*

$$\lim_{n \to \infty} L(F_n, F) = 0.$$

A proof may be found in Lukacs (1975) or in Gnedenko & Kolmogorov (1954).

Next we discuss some properties of these metrics.

Theorem 3.1.2. *Let F and G be two distribution functions. Then the inequality $L(F, G) \leqslant \rho(F, G)$ holds.*

To see this, let $h = \rho(F, G)$. Then for all real x we have the inequalities

$$F(x - h) - h \leqslant G(x) \leqslant F(x + h) + h.$$

Since $h \geqslant 0$ we see that $h \geqslant L(F, G)$.

Theorem 3.1.3. *Let F and G be two distribution functions and let*

$$A = \begin{cases} \sup_x G'(x) & \text{if } G \text{ is absolutely continuous,} \\ \infty & \text{if } G \text{ is not absolutely continuous.} \end{cases}$$

Then $\rho(F, G) \leqslant (1 + A) L(F, G)$.

We write $h = L(F, G)$. Under the assumptions one has either $G(x+h) - G(x) \leqslant hA$ or $G(x-h) - G(x) \geqslant -hA$. In the first case,

$$F(x) \leqslant G(x + h) + h \leqslant G(x) + h(A + 1).$$

In the second case,

$$F(x) \geqslant G(x - h) - h \geqslant G(x) - h(A + 1)$$

so that

$$|F(x) - G(x)| \leqslant h(A + 1)$$

and the statement follows.

Theorem 3.1.4. *Let F_1, F_2, G_1, G_2 be distribution functions. Then*

$$L(F_1 * F_2, G_1 * G_2) \leqslant L(F_1, G_1) + L(F_2, G_2).$$

Let $h_i \geqslant L(F_i, G_i)$ for $i = 1, 2$ and set $h = h_1 + h_2$. For all x and y we have

$$F_1(x - y - h_1) - h_1 \leqslant G_1(x - y) \leqslant F_1(x - y + h_1) + h_1$$

and

$$F_2(x - h_1 - y - h_2) - h_2 \leqslant G_2(x - h_1 - y) \leqslant F_2(x - h_1 - y + h_2) + h_2.$$

We integrate these inequalities with respect to $dG_2(y)$ and $dF_1(y)$ respectively

---

[*] A sequence of distribution functions $F_n(x)$ is said to converge weakly to a distribution function $F(x)$ if it converges in all continuity points of the limiting function. We then write $\lim_{n \to \infty} F_n(x) = F(x)$.

and obtain

(3.1.3) $\quad F_1 * G_2(x - h_1) - h_1 \leqslant G_1 * G_2(x) \leqslant F_1 * G_2(x + h_1) + h_1$

and

(3.1.4) $\quad F_1 * F_2(x - h_2) - h_2 \leqslant F_1 * G_2(x - h_1) \leqslant F_1 * F_2(x - h_1 + h_2) + h_2.$

From this we see that

$$F_1 * F_2(x - h) - h \leqslant F_1 * G_2(x - h_1) - h_1 \leqslant G_1 * G_2(x)$$

and

$$G_1 * G_2(x) \leqslant F_1 * G_2(x + h_1) + h_1 \leqslant F_1 * F_2(x + h) + h.$$

The statement of the theorem follows immediately from the last two inequalities.

The distance of the uniform metric is scale- and translation-invariant,[*] while the Lévy metric is only translation-invariant.

Let $F = F(x)$ and $G = G(x)$ be two distribution functions and write

$$F_a = F(x/a), \quad G_a = G(x/a).$$

Since the Lévy distance is not scale-invariant it is of interest to find a relation between the distance $L(F, G)$ and $L(F_a, G_a)$. J. Thompson (1975) derived the following inequality:

(3.1.5) $\quad \min(a, 1) L(F, G) \leqslant L(F_a, G_a) \leqslant \max(a, 1) L(F, G).$

3.2 Esseen's results and their extension

The principal aim of this section is the derivation of estimates of the distance between two distribution functions in terms of their characteristic functions. The following important results were obtained by C. G. Esseen (1944).

Theorem 3.2.1. *Let A, T and ε be arbitrary positive constants, F(x) a nondecreasing function and G(x) a real function of bounded variation. Let f(t) and g(t) be the Fourier-Stieltjes transforms of F(x) and G(x) respectively. Assume that*

(i) $F(-\infty) = G(-\infty), \quad F(+\infty) = G(+\infty)$
(ii) $G'(x)$ *exists for all x and* $|G'(x)| \leqslant A$

(iii) $\displaystyle\int_{-\infty}^{\infty} |F(x) - G(x)| \, dx < \infty$

(iv) $\displaystyle\int_{-T}^{T} \left| \frac{f(t) - g(t)}{t} \right| dt = \epsilon.$

---

[*] $\rho(x, y)$ is said to be translation-invariant [scale-invariant] if $\rho(x - a, y - a) = \rho(x, y)$ [$\rho(x/a, y/a) = \rho(x, y)$].

Then to every $k>1$ there corresponds a finite, positive $c(k)$ depending only on k, such that

$$|F(x)-G(x)| \leqslant k\frac{\epsilon}{2\pi} + c(k)\frac{A}{T}.$$

Theorem 3.2.2. *Let A, T and ϵ be arbitrary positive constants, $F(x)$ a non-decreasing purely discrete function, and $G(x)$ a real function of bounded variation. Let $f(t)$ and $g(t)$ be the Fourier-Stieltjes transforms of $F(x)$ and $G(x)$ respectively. Assume that*

(i) $F(-\infty) = G(-\infty);\quad F(+\infty) = G(+\infty)$

(ii) $\int_{-\infty}^{\infty} |F(x)-G(x)|\,dx < \infty$

(iii) $\int_{-T}^{T}\left|\frac{f(t)-g(t)}{t}\right|dt = \epsilon$

(iv) *the functions $F(x)$ and $G(x)$ have discontinuities only at the points x_v ($v=0,\pm 1, \pm 2,\ldots; x_{v+1} > x_v$) and there exists a constant $L>0$ such that $\mathrm{Inf}\,(x_{v+1}-x_v) \geqslant L$*

(v) $|G'(x)| \leqslant A$ *for all $x \neq x_v$ ($v=0, \pm 1, \pm 2, \ldots$).*

Then to every number $k>1$ there correspond two finite positive numbers $c_1(k)$ and $c_2(k)$, depending only on k, such that

$$|F(x)-G(x)| \leqslant k\frac{\epsilon}{2\pi} + c_1(k)\frac{A}{T},$$

provided that $TL \geqslant c_2(k)$.

The limit theorems of probability theory give approximations for the distributions of normalized sums of random variables. Esseen's theorems provide an important tool for the study of the error terms of these approximations.

We present here a somewhat generalized version of Theorem 3.2.1; the proof is due to M. Loève (1977).

Theorem 3.2.3. *Let $F(x)$ and $G(x)$ be two distribution functions and write $f(t)$ and $g(t)$ for their characteristic functions. Suppose that $G(x)$ has a derivative $G'(x)$ for all x such that $\sup_x |G'(x)| = A < \infty$. Then*

$$\rho[F,G] = \sup_x |F(x)-G(x)| \leqslant \frac{1}{\pi}\int_{-T}^{T}\left|\frac{f(t)-g(t)}{t}\right|dt + \frac{24A}{\pi T}.$$

ESTIMATION OF THE CLOSENESS OF DISTRIBUTION FUNCTIONS

For the proof of the theorem we need several lemmas. To simplify the notation we write $\Delta = \Delta(x) = F(x) - G(x)$ and $\delta = \delta(t) = f(t) - g(t)$ and put $\alpha = \sup_x |\Delta(x)|$. Without loss of generality we can assume that $\alpha > 0$.

Lemma 3.2.1. *If $G(x)$ is continuous for all x then there exists an x_0 such that either $\Delta(x_0) = \mp \alpha$ or $\Delta(x_0 - 0) = \mp \alpha$.*

Let $\{x_n\}$ be a sequence such that $\lim_{n \to \infty} |\Delta(x_n)| = \alpha$; this sequence contains a convergent subsequence $\{x_{n'}\}$ such that $\lim_{n' \to \infty} x_{n'} = s$. Since $\Delta(\pm \infty) = 0$ and $\alpha > 0$ then necessarily $s < \infty$. The sequence $\{x_{n'}\}$ contains a subsequence $\{x_{n''}\}$ such that either $\lim_{x_{n''} \to \infty} \Delta(x_{n''}) = -\alpha$ or $\lim_{x_{n''} \to \infty} \Delta(x_{n''}) = +\alpha$. We consider only the case $\lim \Delta(x_{n''}) = +\alpha$ since the second case is treated in the same way.

If $\{x_{n''}\}$ contains a subsequence which converges to s from the right, then $+\alpha = \lim_{n'' \to \infty} \Delta(x_{n''}) = \Delta(s)$; hence $x_0 = s$. If $\{x_{n''}\}$ contains a subsequence which converges to s from the left, then

$$\alpha = \lim_{n'' \to \infty} \Delta(x_{n''} - 0) = \Delta(s - 0) = F(s - 0) - G(s - 0).$$

Since G is continuous, we have

$$\alpha = F(s - 0) - G(s) \leq F(s) - G(s) = \Delta(s) = \alpha$$

so that $\Delta(s) = \alpha$ and $s = x_0$.

Let $P(x)$ be an absolutely continuous distribution function with symmetric density $p(x) = p(-x)$ for all x.

Lemma 3.2.2. *Suppose that $G(x)$ has a derivative $G'(x)$ for all x; then there exists a finite real a such that*

$$\left| \int_{-\infty}^{\infty} \Delta(x + a) p(x) \, dx \right| \geq \frac{\alpha}{2} \left\{ 1 - 6 \int_{\gamma}^{\infty} p(x) \, dx \right\}.$$

Here $A = \sup_x |G'(x)|$ and $\gamma = \alpha/(2A)$.

If $A = \infty$ then $\gamma = 0$ and the statement is trivial. We assume therefore that $\gamma > 0$. One has, for any a,

(3.2.1)

$$\left| \int_{-\infty}^{\infty} \Delta(x + a) p(x) \, dx \right| \geq \left| \int_{|x| < \gamma} \Delta(x + a) p(x) \, dx \right| - \left| \int_{|x| \geq \gamma} \Delta(x + a) p(x) \, dx \right|$$

and we note that

$$(3.2.2) \quad \left|\int_{|x|\geq\gamma} \Delta(x+a)\,p(x)\,dx\right| \leq \alpha \int_{|x|\geq\gamma} p(x)\,dx.$$

According to Lemma 3.2.1, there exists an x_0 such that $\Delta(x_0) = -\alpha$. We consider only this possibility since the other three cases are treated in the same way.

Let x be such that $-\gamma < x < \gamma$ and set $a = x_0 - \gamma$. Then $x_0 - 2\gamma < x + a < x_0$, so that $G(x+a) = G(x_0) + \theta(x-\gamma)\,G'(x')$, where $|\theta| \leq 1$ and $x_0 - 2\gamma < x' < x_0$. For $|x| < \gamma$ we have

$$\Delta(x+a) = F(x+a) - G(x_0) - \theta(x-\gamma)\,G'(x')$$

or, since $x - \gamma < 0$,

$$\Delta(x+a) \leq F(x_0) - G(x_0) - A(x-\gamma) \leq -\alpha - A(x-\gamma)$$
$$= -A(2\gamma + x - \gamma) = -A(x+\gamma).$$

Therefore,

$$(3.2.3) \quad \int_{|x|<\gamma} \Delta(x+a)\,p(x)\,dx \leq -A \int_{|x|<\gamma} (x+\gamma)\,p(x)\,dx$$
$$= -A\gamma \int_{|x|<\gamma} p(x)\,dx = -\frac{\alpha}{2}\left[1 - \int_{|x|\geq\gamma} p(x)\,dx\right].$$

We substitute (3.2.3) and (3.2.2) into (3.2.1) and get

$$\left|\int_{-\infty}^{\infty} \Delta(x+a)\,p(x)\,dx\right| \geq \frac{\alpha}{2}\left\{1 - 3\int_{|x|\geq\gamma} p(x)\,dx\right\}$$

and the statement of Lemma 3.2.2 follows immediately.

Let $h(t)$ be a real, absolutely integrable characteristic function. The density function corresponding to $h(t)$ is then given by the inversion formula,

$$p(x) = \frac{1}{2\pi}\int_{-\infty}^{\infty} \cos ux\, h(u)\,du$$

and it is symmetric.

Lemma 3.2.3.

$$\frac{1}{2\pi}\int_{-\infty}^{\infty} \left|\frac{\delta(u)\,h(u)}{u}\right| du \geq \int_{-\infty}^{\infty} \Delta(x+a)\,p(x)\,dx$$

for every real a.

ESTIMATION OF THE CLOSENESS OF DISTRIBUTION FUNCTIONS

It is no restriction to assume that $\delta(u) h(u)/u$ is integrable. The function
$$\Delta_1(x) = \int_{-\infty}^{\infty} \Delta(x-y) p(y) \, dy$$
is the convolution of the functions corresponding to $\delta(u)$ and $h(u)$ and has therefore the Fourier-Stieltjes transform $\delta(u) h(u)$. According to the inversion theorem one has
$$\Delta_1(y) - \Delta_1(x') = \frac{1}{2\pi} \int_{-\infty}^{\infty} \frac{e^{-iuy} - e^{-iux'}}{-iu} \delta(u) h(u) \, du.$$

Since $\lim_{x' \to -\infty} \Delta_1(x') = 0$ and since
$$\lim_{x' \to -\infty} \int_{-\infty}^{\infty} e^{-iux'} \delta(u) h(u)/(-iu) \, du = 0$$
by the Riemann-Lebesgue lemma, we have
$$\int_{-\infty}^{\infty} \Delta(x-y) p(y) \, dy = \frac{1}{2\pi} \int_{-\infty}^{\infty} e^{-iux} \delta(u) h(u)/(-iu) \, du.$$

We replace x by a and y by $-x$ and obtain
$$\int_{-\infty}^{\infty} \Delta(x+a) p(x) \, dx = \frac{1}{2\pi} \int_{-\infty}^{\infty} e^{-iua} \delta(u) h(u)/(-iu) \, du.$$

The statement of Lemma 3.2.3 follows immediately.

We now proceed to the proof of Theorem 3.2.2. We need here the function
$$k(t) = \begin{cases} 1 - |t| & \text{if } |t| \leq 1, \\ 0 & \text{if } |t| \geq 1. \end{cases}$$

$k(t)$ is symmetric and absolutely integrable and
$$p(x) = \frac{1}{2\pi} \int_{-\infty}^{\infty} e^{-itx} k(t) \, dt = \frac{1}{2\pi} \int_{-1}^{1} (1 - |t|) \cos tx \, dt.$$

It is easily seen that $p(x)$ is a density function so that $k(t)$ is a characteristic function. We consider the function $k(t/T)$ ($T > 0$ arbitrary); the corresponding density function $p_T(x)$ is given by
$$p_T(x) = \frac{1 - \cos Tx}{\pi T x^2}.$$

We now replace in Lemmas 3.2.2 and 3.2.3 the function $h(u)$ by $k(u/T)$ and the density $p(x)$ by the density $p_T(x)$. In this way we get

(3.2.4) $$\left| \int_{-\infty}^{\infty} \Delta(x+a) p_T(x) \, dx \right| \geq \frac{\alpha}{2} \left\{ 1 - 6 \int_{\gamma}^{\infty} p_T(x) \, dx \right\}$$

and
(3.2.5) $$\frac{1}{2\pi}\int_{-T}^{T}\left|\frac{\delta(u)}{u}\right|du \geq \frac{1}{2\pi}\int_{-\infty}^{\infty}\left|\frac{\delta(u)\,k(u/T)}{u}\right|du$$
$$\geq \int_{-\infty}^{\infty}\Delta(x+a)\,p_T(x)\,dx.$$

We combine the inequalities (3.2.4) and (3.2.5) and get

(3.2.6) $$\int_{-T}^{T}\left|\frac{\delta(u)}{u}\right|du \geq \frac{\alpha}{2}\left\{1 - 6\int_{\gamma}^{\infty}p_T(x)\,dx\right\}.$$

Here
$$\int_{\gamma}^{\infty}p_T(x)\,dx = \frac{1}{\pi}\int_{\gamma}^{\infty}\frac{1-\cos Tx}{Tx^2}\,dx \leq \frac{2}{\pi}\int_{\gamma}^{\infty}\frac{dx}{Tx^2} = \frac{2}{\pi\gamma T} = \frac{4A}{\pi\alpha T},$$

so that
$$\frac{1}{2\pi}\int_{-T}^{T}\left|\frac{f(u)-g(u)}{u}\right|du \geq \frac{\alpha}{2} - \frac{12A}{\pi T}$$

and the statement of the theorem follows immediately.

The statement of the theorem can be written as

$$\sup_{x}|F(x)-G(x)| \leq \frac{2}{\pi}\int_{0}^{T}\left|\frac{f(t)-g(t)}{t}\right|dt + \frac{24A}{\pi T}.$$

If $T \to \infty$, this becomes

(3.2.7) $$\rho(F,G) = \sup_{x}|F(x)-G(x)| \leq \frac{2}{\pi}\int_{0}^{\infty}|f(t)-g(t)|\frac{dt}{t}.$$

3.3 Distances in the Lévy metric

For one-sided distribution functions G. Laue obtained a generalization of Theorem 3.2.3.

Next we consider the Lévy metric in order to estimate the distance in this metric in terms of characteristic functions.

Theorem 3.3.1. *Let $F(x)$ and $G(x)$ be two distribution functions with characteristic functions $f(t)$ and $g(t)$ respectively. Then*

$$L(F,G) \leq \frac{2}{\pi}\int_{0}^{T}|f(t)-g(t)|\frac{dt}{t} + 2\,e\,\frac{\log T}{T},$$

provided that $T \geq 1\cdot 3$.

A slightly more general version of this theorem was established by V. M. Zolotarev (1970, 1971).

For the proof of the theorem we need the following lemma.

Lemma 3.3.1. *Let $F(x)$, $G(x)$ and $H(x)$ be three distribution functions; then the inequalities*

$$0 \leqslant L(F, G) - L(F*H, G*H)$$
$$\leqslant \max(2\epsilon, 1 - H(\epsilon) + H(-\epsilon))$$

hold for any $\epsilon > 0$.

For the sake of brevity we put $L' = L(F, G)$ and $L'' = L(F*H, G*H)$. We also write $G*H(x)$ for the convolution $\int_{-\infty}^{\infty} G(x-z)\,dH(z)$ if we wish to put the variable into evidence.

For any $L_0 > L'$ one has

$$G(x - L_0) - L_0 \leqslant F(x) \leqslant G(x + L_0) + L_0.$$

We multiply these inequalities by $-dH(y-x)$ and integrating them over all x we see that

$$G*H(y - L_0) - L_0 \leqslant F*H(y) \leqslant G*H(y + L_0) + L_0$$

for all y. Therefore $L'' \leqslant L_0$. We let L_0 tend to L' and obtain the first inequality of the lemma.

We have

$$F*H(x) = \int_{-\infty}^{\epsilon} F(x-y)\,dH(y) + \int_{\epsilon}^{\infty} F(x-y)\,dH(y)$$

$$\geqslant F(x-\epsilon)H(\epsilon) - \int_{\epsilon}^{\infty} F(x-y)\,d[1-H(y)].$$

Therefore $F*H(x) \geqslant F(x-\epsilon) - 1 + H(\epsilon)$.

In a similar way we obtain $F*H(x) \leqslant H(-\epsilon) + F(x+\epsilon)$. We replace x by $x + \epsilon$ and see that

$$F(x) - 1 + H(\epsilon) \leqslant F*H(x+\epsilon)$$

and similarly

$$G(x) - 1 + H(\epsilon) \leqslant G*H(x+\epsilon).$$

Therefore

$$G(x) \leqslant G*H(x+\epsilon) + 1 - H(\epsilon)$$
$$\leqslant F(x + 2\epsilon + L'') + L'' + 1 - H(\epsilon) + H(-\epsilon),$$

so that

$$G(x) \leqslant F(x + 2\epsilon + L'') + L'' + 1 - H(\epsilon) + H(-\epsilon).$$

Similarly we see that

$$G(x) \geq F(x - 2\epsilon - L'') - L'' - 1 + H(\epsilon) - H(-\epsilon)$$

and conclude easily that

$$L' = L(F, G) \leq \max(2\epsilon, 1 - H(\epsilon) + H(-\epsilon)) + L''$$

and Lemma 3.3.1 is proved.

We introduce the distribution function $K_\epsilon(x)$ which is defined by its characteristic function

$$(3.3.1) \quad k_\epsilon(t) = \left[\frac{\sin \epsilon t/2n}{\epsilon t/2n}\right]^n.$$

The positive integer n and $\epsilon > 0$ will be chosen later.

The function $\dfrac{\sin \epsilon t/2n}{\epsilon t/2n}$ is the characteristic function of a rectangular distribution over $(-\epsilon/2n, \epsilon/2n)$; therefore $K_\epsilon(x)$ has as support the interval $(-\epsilon/2, \epsilon/2)$, that is, $1 - K_\epsilon(x) = K_\epsilon(-x) = 0$ for $|x| > \epsilon/2$. It follows from Lemma 3.3.1 that

$$(3.3.2) \quad L(F, G) \leq \epsilon + L(F * K_\epsilon, G * K_\epsilon).$$

According to Theorem 3.1.2, $L(F, G) \leq \rho(F, G)$ and we see from formula (3.2.7) that

$$L(F * K_\epsilon, G * K_\epsilon) \leq \rho(F * K_\epsilon, G * K_\epsilon)$$
$$\leq \frac{2}{\pi} \int_0^\infty |f(t) - g(t)| |k_\epsilon(t)| \frac{dt}{t} = J.$$

Let $T \geq \pi e$, and estimate the integral J. We have

$$J \leq \frac{2}{\pi} \int_0^T |f(t) - g(t)| \frac{dt}{t} + \frac{4}{\pi} \int_T^\infty |k_\epsilon(t)| \frac{dt}{t} = J_1 + J_2 \text{ (say)}.$$

Then we conclude from (3.3.1) that

$$J_2 \leq \frac{4}{\pi} \int_T^\infty \left[\frac{\sin \epsilon t/2n}{\epsilon t/2n}\right]^n \frac{dt}{t} \leq \frac{4}{n\pi} \left(\frac{2n}{T\epsilon}\right)^n.$$

Putting $B = \dfrac{4}{\pi}\left(\dfrac{2n}{T}\right)^n$, one has

$$J_2 \leq \frac{1}{n} B \epsilon^{-n}.$$

We now choose $\epsilon = B^{1/(1+n)}$ and we get from (3.3.2)

ESTIMATION OF THE CLOSENESS OF DISTRIBUTION FUNCTIONS

(3.3.3) $\quad L(F, G) \leqslant J_1 + \left(1 + \dfrac{1}{n}\right) B^{1/(1+n)}.$

It is then easily seen that

(3.3.4) $\quad \dfrac{n+1}{n} B^{1/(1+n)} < 2^{[1+1/(n+1)]} \dfrac{(n+1)}{T} \left(\dfrac{T}{\pi n}\right)^{1/(1+n)}$

$$< 4 \dfrac{(n+1)}{T} \left(\dfrac{T}{\pi n}\right)^{1/(1+n)}$$

Finally we select n so that

$$n \leqslant \log(T/\pi) < n+1.$$

Then

(3.3.5) $\quad \left(\dfrac{T}{n\pi}\right)^{1/(n+1)} \leqslant (T/\pi)^{1/(n+1)} < e$

and it follows from (3.3.4) that

(3.3.6) $\quad \dfrac{n+1}{n} B^{1/(1+n)} < \dfrac{4e}{T}\left(1 + \log \dfrac{T}{\pi}\right).$

We conclude from the definition of J_1, (3.3.3) and (3.3.6), that

$$L(F, G) \leqslant \dfrac{2}{\pi} \int_0^T |f(t) - g(t)| \dfrac{dt}{t} + \dfrac{4e}{T}\left(1 + \log \dfrac{T}{\pi}\right).$$

Since $1 + \log \dfrac{T}{\pi} = \log T + \log \dfrac{e}{\pi} < \log T$, we can write

(3.3.7) $\quad L(F, G) \leqslant \dfrac{2}{\pi} \int_0^T |f(t) - g(t)| \dfrac{dt}{t} + 4e \dfrac{\log T}{T},$

provided that $T \geqslant \pi e$. However, one easily sees that $2e \log T/(T) \geqslant 1$ for $1 \cdot 3 \leqslant T < \pi e$, so that the last estimate for $L(F, G)$ is also valid for $T \geqslant 1 \cdot 3$.

Corollary to Theorem 3.3.1. *Under the assumptions of Theorem 3.3.1 one has*

$$L(F, G) \leqslant \dfrac{2}{\pi} \int_0^\infty |f(t) - g(t)| \dfrac{dt}{t}.$$

provided that this integral exists. The statement of the corollary is obtained by letting T tend to ∞ in formula (3.3.7).

For the proof of the next theorem we need the distribution function $K(x)$, defined by its characteristic function[*]

(3.3.8) $\quad k(t) = \left[\dfrac{\sin \eta t/2m}{\eta t/2m}\right]^m \quad (0 < \eta < 1).$

Theorem 3.3.2. *Let F and G be two distribution functions with characteristic functions $f(t)$ and $g(t)$ respectively, and let $r > 0$ be a real constant. Then*

$$L(F, G) \leqslant (rCD)^{1/(1+r)} \left(1 + \dfrac{1}{r}\right).$$

Here

$$C = \dfrac{2}{\pi} \int_0^\infty |k(t)| \, t^{r-1} \, dt \quad \text{while}$$

$$D = \sup_{t > 0} t^{-r} |f(t) - g(t)| \quad \text{and} \quad \pi rC/2 \leqslant (2r+2)^{r+1}.$$

Formula (3.3.2) is again valid, so that

(3.3.9) $\quad L(F, G) \leqslant \eta + L(F * K, G * K).$

Moreover we see from the corollary to Theorem 3.3.1 that

$$L(F * K, G * K) \leqslant \dfrac{2}{\pi} \int_0^\infty |f(t) - g(t)| |k(t)| \, \dfrac{dt}{t} = I, \quad \text{(say)}.$$

We write I in the form

$$I = \dfrac{2}{\pi} \int_0^\infty (t\eta)^{-r} |f(t) - g(t)| \cdot |k(t)| \dfrac{dt}{\eta^{-r} t^{-r+1}}.$$

Therefore

$$I \leqslant \{\sup_{t > 0} [(t\eta)^{-r} |f(t) - g(t)|]\} \dfrac{2}{\pi} \int_0^\infty \dfrac{|k(t)| \, dt}{\eta^{-r} t^{-r+1}}$$

or

(3.3.10) $\quad I \leqslant \eta^{-r} DC.$

It follows from (3.3.9) and (3.3.10) that

$$L(F, G) \leqslant \eta + \eta^{-r} DC.$$

---

[*] This is essentially the same function as we used before [see formula (3.3.1)]. We changed the notation in order to emphasize that the parameters η and m of (3.3.8) will be chosen in a different way.

ESTIMATION OF THE CLOSENESS OF DISTRIBUTION FUNCTIONS 43

The right-hand side of this inequality is minimized by putting $\eta = (rDC)^{1/(r+1)}$. With this value for η we obtain

$$L(F, G) \leq \left(1 + \frac{1}{r}\right)(rDC)^{1/(1+r)}.$$

Corollary to Theorem 3.3.2. Under the conditions of the theorem one can show that for a suitably chosen m the relation

$$\pi rC/2 \leq (2r + 2)^{r+1}$$

holds

We select m as the integer nearest to $r + 1$; it is then easy to show that

$$\frac{m^{r+1}}{2(m-r)} \leq (r+1)^{r+1}$$

or equivalently

$$\frac{(2m)^{r+1}}{2(m-r)} \leq (2r+2)^{r+1}.$$

It follows from the definition of C (see the statement of Theorem 3.3.2) that

$$\pi rC/2 = r \int_0^\infty |k(t)| t^{r-1} \, dt$$

$$= r \int_0^{2m} |k(t)| t^{r-1} \, dt + r \int_{2m}^\infty |k(t)| t^{r-1} \, dt$$

$$\leq (2m)^r + \frac{r}{m-r}(2m)^r = \frac{(2m)^{r+1}}{2(m-r)} \leq (2r+2)^{r+1}.$$

Therefore

$$\pi rC/2 \leq (2r+2)^{r+1}$$

and the proof of the corollary to Theorem 3.3.2 is completed.

A number of authors have discussed variants of these theorems. We mention only F. J. Dyson (1953), M. Motoo (1955), L. D. Meshalkin & B. A. Rogozin (1963), V. M. Zolotarev (1965, 1967), A. S. Fainleib (1968), V. I. Pauluskas (1971), W. Feller (1971), J. L. Paditz (1975) and V. A. Abramov (1976).

We indicate briefly some of the results obtained by Meshalkin & Rogozin (1963) who replaced the integral

$$\int_{-T}^{T} \left|\frac{f(t) - g(t)}{t}\right| dt$$

used in Theorem 3.3.1 by $\sup_{|t|\leqslant T} |f(t) - g(t)|$ and obtained the following:

Theorem 3.3.3. Let $F(x)$ be a bounded, non-decreasing function and let $G(x)$ be a function of bounded variation. Denote the Fourier-Stieltjes transforms of $F(x)$ and $G(x)$ respectively by $f(t)$ and $g(t)$. Suppose that

(i) $F(-\infty) = G(-\infty)$,
(ii) $G'(x)$ *exists for all x and* $|G'(x)| \leqslant A$,
(iii) $|f(t) - g(t)| < \epsilon$ *for* $|t| < T$,

where A, T and ϵ are positive numbers. Then for any $L > 2/T$ one has

$$|F(x) - G(x)| < C\left[\epsilon \log(LT) + \frac{A}{T} + \gamma(L)\right].$$

Here C is an absolute constant and

$$\gamma(L) = \underset{-\infty < x < \infty}{\mathrm{var}} G(x) - \sup_x \underset{x \leqslant y \leqslant x+L}{\mathrm{var}} G(y)$$

and $\underset{a \leqslant x \leqslant b}{\mathrm{var}}\, G(x)$ denotes the total variation of $G(x)$ in the interval $[a, b]$.

We also note that assumption (ii) of Theorem 3.2.1 is not needed here.

4 Extensions of the concept of stable distributions

4.1 Generalizations of stable distributions

The concept of stable distributions (or of stable characteristic functions) admits several generalizations. These wider classes of characteristic functions are, as the stable characteristic functions, defined by means of functional equations which they satisfy.

P. Lévy (1937a) had already introduced the first extension by defining semi-stable distributions as distribution functions whose characteristic functions satisfy, for all t, the equation

(4.1.1) $\quad f(t) = [f(\beta t)]^{\gamma}$,

where $0 < |\beta| < 1, \gamma > 0$.

The solutions of (4.1.1) admit the canonical representation [see Kagan, Linnik & Rao (1973)]

(4.1.2) $\quad \log f(t) = iat + \int_{-\infty}^{0-} \left(e^{itx} - 1 - \frac{itx}{1+x^2}\right) dM(x)$

$\qquad + \int_{0+}^{\infty} \left(e^{itx} - 1 - \frac{itx}{1+x^2}\right) dN(x),$

where $M(x)$ and $N(x)$ have the form

$$M(x) = \xi_1(\log|x|)/|x|^{\lambda}, \quad N(x) = -\xi_2(\log x)/x^{\lambda}.$$

Here λ is the unique real solution of the equation $\gamma|\beta|^{\lambda} = 1$, and ξ_1 and ξ_2 are non-negative right continuous functions defined on R^1 which are periodic with period $-\log|\beta|$.

Theorem 4.1.1. *All semi-stable distributions are infinitely divisible.*

This follows immediately from formula (4.1.2).

Ramachandran and Rao (1968) studied farther-reaching generalization of stable distribution functions.

A characteristic function $f(t)$ is said to belong to a generalized stable distribution if it is non-vanishing and if it satisfies the functional equation

(4.1.3) $\quad \displaystyle\prod_{j=1}^{s} [f(c_j t)]^{\gamma_j} = \prod_{j=s+1}^{s+k} [f(c_j t)]^{\gamma_j}$

for all t. Here $\gamma_j > 0$ and $0 < |c_j| \leq 1$ for $1 \leq j \leq s + k$. Two special cases are of interest:

$$(4.1.3a) \quad f(t) = \prod_{j=1}^{r} [f(c_j t)]^{\gamma_j}$$

with $\gamma_j > 0$ and $0 < |c_j| < 1$, and

$$(4.1.3b) \quad f(t) = \prod_{j=1}^{r} [f(c_j t)]^{\gamma_j},$$

with $\gamma_j > 0$ and $0 < c_j < 1$,

For $r = 1$, equation (4.1.3b) becomes the equation defining semi-stable distributions.

Let λ be the unique real solution of the equation

$$(4.1.4) \quad \sum_j \gamma_j c_j^\lambda = 1.$$

Ramachandran and Rao obtained a number of interesting results. We formulate here only part of their results.

Theorem 4.1.2. *Suppose that either* $0 < \lambda < 1$ *or* $1 < \lambda < 2$; *then the non-degenerate solutions of* (4.1.3b) *belong to absolutely continuous, infinitely divisible distributions. They have moments of all positive orders inferior to* λ.

Theorem 4.1.3. *Let* $f(t)$ *be the characteristic function of a generalized stable distribution* [*i.e. the characteristic function satisfying* (4.1.3a)]; *then* $f(t)$ *is infinitely divisible. If* λ *is the solution of* (4.1.4), *then the corresponding distribution function has moments of all positive orders inferior to* λ *but no moments of order greater than or equal to* λ. *If* $\lambda = 2$ *then the distribution is normal.*

Remark. In the discussion of (4.1.3b) we omitted the cases where the equation yields degenerate solutions.

A distribution function $G(x)$ is said to belong to the domain of attraction of a distribution function $F(x)$ if for some sequences of constants $\{A_n\}$ and $\{B_n\}$,

$$\lim_{n \to \infty} G^{*n}(B_n x + A_n) = F(x).$$

R. Shimizu (1970) gave a necessary and sufficient condition which ensures that a distribution function $G(x)$ belongs to the domain of attraction of a semi-stable distribution function $F(x)$.

4.2 Self-decomposable distributions

A characteristic function $f(t)$ is said to be self-decomposable if it satisfies the relation

$$f(t) = f(ct) f_c(t)$$

for every c ($0 < c < 1$) where $f_c(t)$ is some characteristic function.

Distribution functions which belong to self-decomposable characteristic functions are called self-decomposable distribution functions. Some authors refer to this family as the L-class.

Some classical results on self-decomposable distributions can be found in Lukacs (1970). We quote here one of these which will be used in the sequel.

Theorem 4.2.1. An infinitely divisible characteristic function $f(t)$ is self-decomposable if, and only if, the functions $M(u)$ and $N(u)$ in its Lévy canonical representation (Theorem 1.3.4) have left and right derivatives everywhere, and if the function $uM'(u)$ is non-decreasing for $u < 0$ while $uN'(u)$ is non-increasing for $u > 0$. Here $M'(u)$ and $N'(u)$ denote either right- or left-hand derivatives, possibly different ones at different points.

The proof of this theorem is given in Lukacs (1970).

From Theorem 4.2.1 it follows that the characteristic function $f(t)$ of a self-decomposable distribution whose spectral function $N(u) = 0$ can be written in the form [see Wolfe (1971a)]

$$\log f(t) = i\gamma t - \sigma^2 t^2/2 + \int_{+0}^{+\infty} \left(e^{itu} - 1 - \frac{itu}{1+u^2} \right) u^{-1} \lambda(u) \, du,$$

where $\lambda(u) = uM'(u)$ is non-decreasing on $(0, \infty)$. A similar statement can be made concerning the characteristic function of a self-decomposable distribution with $M(u) = 0$. Thus one obtains the following representation for characteristic functions of self-decomposable distributions:

$$(4.2.1) \quad \log f(t) = i\gamma t - \sigma^2 t^2/2 + \int_{-\infty}^{-0} \left(e^{itu} - 1 - \frac{itu}{1+u^2} \right) |u|^{-1} \mu(u) \, du$$

$$+ \int_{+0}^{+\infty} \left(e^{itu} - 1 - \frac{itu}{1+u^2} \right) u^{-1} \lambda(u) \, du,$$

where the functions $\mu(u)$ and $\lambda(u)$ are non-negative and satisfy certain conditions imposed by the properties of spectral functions.

Corollary to Theorem 4.2.1. All stable characteristic functions are self-decomposable.

The corollary follows immediately from Theorem 4.2.1.

K. Urbanik (1968, 1973) gave the following representation for self-decomposable characteristic functions $f(t)$.

Theorem 4.2.2. A self-decomposable characteristic function $f(t)$ admits the representation

$$\log f(t) = iat + \int_{-\infty}^{\infty} \left[\int_{0}^{\mu} \frac{e^{iv}-1}{v} \, dv - it \arctan u \right] \frac{dQ(u)}{\log(1+u^2)}.$$

Here a is a real constant, Q is a finite Borel measure on R_1; the integrand is defined for $u = 0$ by continuity to be $-t^2/4$. Moreover, a and $Q(u)$ are uniquely determined by $f(t)$. A detailed proof of this theorem can be found in Urbanik (1975).

It is also possible to define a sequence of classes $\{L_m\}$ of characteristic functions (distribution functions).

A characteristic function $f(t)$ is said to belong to the class L_m if, for every a, $0 < a < 1$, there exists a characteristic function $f_a(t) \in L_{m-1}$ such that

(4.2.2) $\quad f(t) = f(at) f_a(t)$.

The class L_{-1} is by definition the class of all characteristic functions; the class $L_0 = L$ is the class of self-decomposable characteristic functions.

Urbanik (1973) and also A. Kumar and B. M. Schreiber (1978) studied the classes L_m and they obtained representations similar to the representation of characteristic functions from L. The classes L_m form a sequence of decreasing sets, they are closed under shifts, changes of scale, convolutions and weak convergence.

Let $f(t)$ be an arbitrary characteristic function. Denote by $\theta(f) = \{m : m \geq 0, f(t) \in L_m\}$. These classes were also studied by K. Sato (1978) and by A. I. Ilinskii (1979).

V. M. Zolotarev (1963) investigated the smoothness properties of self-decomposable distribution functions. In this paper he also showed that self-decomposable distributions are absolutely continuous. Recently the concepts of self-decomposability and of stability have been extended to discrete distributions on the non-negative integers by F. W. Steutel and K. van Harn (1979).

5 Unimodality

A distribution function $F(x)$ is said to be unimodal if there exists at least one point a such that $F(x)$ is convex for $x < a$ but concave for $x > a$. Such a point is called a vertex (or mode) of $F(x)$.

In this chapter we study unimodal distribution functions, that is, distributions which have exactly one mode. Examples of unimodal distributions are the Normal and the Cauchy distributions.

5.1 Conditions for unimodality

We mention first a result due to A. Ya. Khinchine.

Theorem 5.1.1. *A distribution function $F(x)$ is unimodal with mode at the origin if and only if its characteristic function can be represented as*

$$f(t) = \frac{1}{t}\int_0^t g(u)\, du,$$

where $g(u)$ is a characteristic function.

The next theorem applies to absolutely continuous, symmetric distributions.

Theorem 5.1.2. *Let $f(t)$ be a continuous, real-valued function such that $f(-t) = f(t)$ and $f(0) = 1$. Let $A(z)$ be a function of $z = t + iy = re^{i\theta}$ (t, y, θ, r real) which satisfies the following conditions:*

  (i) *$A(z)$ is regular in the region*

$$\mathscr{D} = \left\{ z : r > 0,\ -\epsilon_1 < \theta < \frac{\pi}{2} + \epsilon_2 \right\},$$

   where $\epsilon_1, \epsilon_2 > 0$ can be arbitrarily small.
  (ii) $|A(z)| = O(1)$ as $|z| \to 0$
  (iii) $|A(z)| = O(|z|^{-\delta})$ as $|z| \to \infty$ $(\delta > 1)$
  (iv) $\operatorname{Im} A(iy) \leq 0$ for $y > 0$
  (v) $f(t) = A(t)$ for $t > 0$

Then $f(t)$ is the characteristic function of a symmetric, unimodal and absolutely continuous distribution function.

Another important result is the following.

Theorem 5.1.3. Let $\{F_n(x)\}$ be a sequence of distribution functions which converges weakly to a distribution function $F(x)$. Suppose the $F_n(x)$ are unimodal, then $F(x)$ is also unimodal.

Theorem 5.1.4. Let $F(x)$ be a unimodal distribution function. Then there exists a sequence of distribution functions $\{F_n(x)\}$ such that $F_n(x)$ is absolutely continuous and unimodal and that $\lim_{n\to\infty} F_n(x) = F(x)$ and $F'_n(x)$ is absolutely continuous.

Theorem 5.1.5. Let α be a positive real number $\alpha \leq 2$, then $f(t) = \dfrac{1}{1+|t|^\alpha}$ is the characteristic function of a unimodal distribution.

For the proof of the last two theorems we refer to Lukacs (1970).

5.2 Convolutions of unimodal distributions

For a long time it was believed that the convolution of two unimodal distributions is unimodal, but this belief is not justified. It originated in a statement given in the thesis of A. I. Lapin (1947) and was used by B. V. Gnedenko and A. N. Kolmogorov (1954) to prove the unimodality of all stable distributions. Since Lapin's proof is incorrect, this proof of the unimodality of all stable distributions is also invalid and had to be replaced by an entirely different argument (see Section 5.4). It was K. L. Chung who showed that Lapin's proof is incorrect by constructing a counter-example which we present next.

Chung (1953) gave the following example of an absolutely continuous distribution function $F(x)$ which is unimodal with vertex $x = 0$ but which has the property that the density function of $F*F$ has two maxima, so that $F*F$ is not unimodal. Let

$$p_1(x) = \begin{cases} 0 & \text{if } x < -\tfrac{1}{30} \text{ or if } x > \tfrac{5}{6} \\ 5 & \text{if } -\tfrac{1}{30} \leq x \leq 0 \\ 1 & \text{if } 0 < x \leq \tfrac{5}{6} \end{cases}$$

be the density function of $F(x)$ and write $p_2(x)$ for the density function of $F*F$. Then

$$p_2(x) = \begin{cases} 0 & \text{if } x \leq -\tfrac{1}{15} \text{ or if } x \geq \tfrac{5}{3} \\ 25x + \tfrac{5}{3} & \text{if } -\tfrac{1}{15} \leq x \leq -\tfrac{1}{30} \\ -\tfrac{1}{15}x + \tfrac{1}{3} & \text{if } -\tfrac{1}{30} \leq x \leq 0 \\ x + \tfrac{1}{3} & \text{if } 0 \leq x \leq \tfrac{4}{5} \\ -9x + \tfrac{25}{3} & \text{if } \tfrac{4}{5} \leq x \leq \tfrac{5}{6} \\ \tfrac{5}{3} - x & \text{if } \tfrac{5}{6} \leq x \leq \tfrac{5}{3} \end{cases}$$

UNIMODALITY

The function $p_2(x)$ is continuous and has two maxima at the points $x = -\frac{1}{30}$ and $x = \frac{5}{4}$ and a minimum at the point $x = 0$. Therefore $F * F$ is not unimodal.

We next discuss a sufficient condition which ensures that a characteristic function belongs to a unimodal distribution. This condition, as well as its proof, is similar to Pólya's Theorem 1.2.3.

Theorem 5.2.1. *Let $f(t)$ be a real-valued continuous function such that $f(0) = 1$, $f(-t) = f(t)$, $\lim_{t \to \infty} f(t) = 0$ and suppose that $-f'(t)$ is convex for $t > 0$. Assume further that $-f'''(t)$ exists and is non-negative for $t \geq 0$ and that*

$$\int_0^\infty t|f(t)|\, dt \text{ is finite while}$$

$$\lim_{t \to \infty} t^2 f(t) = \lim_{t \to \infty} t^3 f'(t) = \lim_{t \to \infty} t^4 (f''(t)) = 0.$$

Then $f(t)$ is the characteristic function of a unimodal distribution.

Let $k(t) = \int_{-\infty}^\infty f(x) e^{itx}\, dx = 2 \int_0^\infty f(x) \cos xt\, dx$. Clearly $k(t)$ is an even function; therefore we must show that $k'(t) < 0$ for $t > 0$. Since

$$k'(t) = -2 \int_0^\infty x f(x) \sin xt\, dx$$

we have to show that

$$I(t) = \int_0^\infty x f(x) \sin xt\, dx > 0 \text{ for } t > 0.$$

We integrate by parts and use our assumptions and get

$$I(t) = f(x) \int_0^\infty u \sin tu\, du \Big|_0^\infty - \int_0^\infty f'(t) \left[\int_0^\infty u \sin tu\, du \right] dx.$$

We repeat this procedure twice and obtain

$$I(t) = \int_0^\infty f''(x) \int_0^x (x-u) u \sin tu\, du\, dx$$

$$= -\tfrac{1}{2} \int_0^\infty f'''(x) \int_0^x (x-u)^2 u \sin tu\, du\, dx$$

By assumption, $-f'''(x) > 0$ for $x > 0$; therefore we have to show that

(5.2.1) $\quad g(x) = \int_0^x (x-u)^2 u \sin tu\, du > 0$

for $x>0$, $t>0$. We write $v = tu$ so that

$$g(x) = \frac{1}{t^4} \int_0^{xt} (xt-v)^2 \, v \sin v \, dv.$$

Hence it is sufficient to prove that

$$h(x) = \int_0^x (x-v)^2 \, v \sin v \, dv > 0 \text{ for } x > 0.$$

The function $h(x)$ is a convolution, and its Laplace transform is

$$H(z) = \int_0^\infty e^{-xz} \int_0^x (x-v)^2 \, v \sin v \, dv \, dx$$

$$= \int_0^\infty e^{-xz} x^2 \, dx \int_0^\infty e^{-zv} \, v \sin v \, dv$$

$$= \frac{2}{z^3} \operatorname{Im} \int_0^\infty e^{-(z-i)v} \, v \, dv$$

$$= \frac{2}{z^3} \operatorname{Im} \frac{1}{(z-i)^2} = \frac{2 \cdot 2z}{z^3(z^2+1)^2} = \frac{4}{z^2(z^2+1)^2}.$$

Since $\dfrac{1}{z(z^2+1)} = \int_0^\infty e^{-xz}(1-\cos x)\, dx$, we see that

$$\int_0^x (x-v)^2 \, v \sin v \, dv = 4 \int_0^x (1-\cos v)[1-\cos(x-v)] \, dv$$

and this is positive for $x > 0$.

R. Askey (1975) also proved a stronger theorem using Theorem 4.2.1 by reducing its assumption.

Theorem 5.2.2. Let $f(t)$ be a real-valued continuous function such that $f(0) = 1$, $f(-t) = f(t)$, $\lim_{t \to \infty} f(t) = 0$ and suppose that $-f'(t)$ is convex for $t > 0$. Then $f(t)$ is the characteristic function of a unimodal distribution.

Example. Let $f_a(t) = \exp[-a|t| - |t|^3]$. Using Theorem 5.2.2, one can show that $f_a(t)$ is the characteristic function of a unimodal distribution, provided that $a \geq 3$.

A. Olshin and L. J. Savage (1970) generalized the concept of unimodality. Let $G(z)$ be a distribution function which is absolutely continuous and whose density is $p(z)$. $G(x)$ is said to be α-unimodal about the point v if

$t^\alpha \int_{-\infty}^{\infty} f[t(z-v)]\,dG(z)$ is non-decreasing for every bounded measurable function $f(t)$ for $z > 0$. For $\alpha = 1$ this becomes the unimodality as defined at the beginning of Chapter 5. They also showed that the convolution of an α-unimodal and a β-unimodal distribution is $(\alpha + \beta)$-unimodal.

S. W. Dharmadhikari and K. Jogdeo (1974) studied the concept of α-unimodality and obtained the following result:

Theorem 5.2.3. Let F_1 and F_2 be two univariate distributions and assume that F_1 is symmetric and unimodal about zero while F_2 is α-unimodal about zero. Then

(a) If $\frac{1}{2} \leq \alpha \leq 1$ then $F_1 * F_2$ is $\frac{3}{2}$-unimodal about zero.
(b) If $1 \leq \alpha \leq 2$, then $F_1 * F_2$ is $(2 + \alpha)/2$-unimodal about zero.
(c) If $\alpha \geq 2$, then $F_1 * F_2$ is α-unimodal about zero.

Remark. The assumption that one of the factors of the convolution is symmetrical is essential for the validity of the theorem.

While many authors have studied the unimodality of absolutely continuous distributions, little work has been done on purely discrete distributions.

A purely discrete distribution $\{p_u\}$ ($u = \ldots -1, 0, 1, 2, \ldots$) is said to be unimodal if in the sequence

$$\ldots p_{-2} - p_{-1}, p_0 - p_{-1}, p_1 - p_0, \ldots, p_u - p_{u-1} \ldots$$

only one change of sign occurs.

Medgyessy (1972) as well as Dharmadhikari and Jogdeo (1974) have studied discrete unimodal distributions.

5.3 Strong unimodality

I. A. Ibragimov (1956) calls a distribution function strongly unimodal if it is unimodal and if its convolution with any unimodal distribution is unimodal. He obtained the following result:

Theorem 5.3.1. The normal distribution is strongly unimodal.

We consider first unimodal distribution functions $F(x)$ which are absolutely continuous and have an absolutely continuous derivative $F'(x)$. Let $F(x)$ be such a distribution function and consider the distribution function $G(x) = F(x) * \Phi(x\sigma)$, where $\Phi(x)$ is the normal distribution with zero mean and unit variance. Then

$$G''(x) = \frac{\sigma}{\sqrt{2\pi}} \int_{-\infty}^{\infty} \frac{d}{dt} \left\{ \exp -\frac{\sigma^2}{2}(x-t)^2 \right\} F'(t)\,dt$$

$$= \frac{\sigma}{\sqrt{2\pi}} \int_{-\infty}^{\infty} \exp\{-\sigma^2(x-t)^2/2\}\, F''(t)\,dt.$$

If $t>0$ then $F''(t) \leq 0$ and $F''(-t) \geq 0$. We have to show that $G''(x_0) = 0$ implies that $G''(x_0 + \delta) \leq 0$ for $\delta > 0$. We have

$$G''(x_0) = \frac{\sigma}{\sqrt{2\pi}} \left\{ \int_0^\infty F''(t) \exp[-\sigma^2(x_0-t)^2/2] \, dt \right.$$

$$\left. + \int_0^\infty F''(-t) \exp[-\sigma^2(x_0+t)^2/2] \, dt \right\} = 0$$

$$G''(x_0+\delta) = \frac{\sigma}{\sqrt{2\pi}} \exp[(-\sigma^2\delta^2/2)(-\sigma^2\delta x_0)] \left\{ \int_0^\infty F''(t) \exp[-\sigma^2(x_0-t)^2/2] \right.$$

$$\left. \times e^{\sigma^2\delta t} \, dt + \int_0^\infty F''(-t) \exp[-\sigma^2(x_0+t)^2/2] \, e^{-\sigma^2\delta t} \, dt \right\}$$

or

$$G''(x_0+\delta) = \frac{\sigma}{\sqrt{2\pi}} \exp\left[-\frac{\sigma^2\delta^2}{2} - \sigma^2\delta x_0\right]$$

$$\times \left\{ e^{\sigma^2\delta \xi_1} \int_0^\infty F''(t) \exp[-\sigma^2(x_0-t)^2/2] \, dt \right.$$

$$\left. + e^{-\sigma^2\delta \xi_2} \int_0^\infty F''(-t) \exp[-\sigma^2(x_0+t)^2/2] \, dt \right\}$$

so that

$$G''(x_0+\delta) \leq 0 \quad (\delta > 0).$$

This follows from the fact that $A + B = 0$ with $A \leq 0, B \geq 0$, and $a \geq b$ implies that $aA + bB = (b-a)B \leq 0$.

The strong unimodality of $\Phi(x\sigma)$ follows from Theorems 5.1.3 and 5.1.4.

Ibragimov (1956) obtained the following necessary and sufficient condition for strong unimodality.

Theorem 5.3.2. A non-degenerate unimodal distribution function $F(x)$ is strongly unimodal if, and only if, $F(x)$ is continuous and if $\log F'(x)$ is concave on the set E of points on which neither the right-hand nor the left-hand derivative of $F(x)$ vanishes.

We give here only an indication of the proof and refer the reader to Ibragimov (1956) for details. The proof of Theorem 5.3.2 uses at one point Theorem 5.3.1.

First we prove the necessity of the condition under the restrictive assumption

(A) $F'(x)$ exists and is continuous while $E = (-\infty, \infty)$.

We give an indirect proof and suppose that $\psi(x) = \log F'(x)$ is not concave. Then there exists a positive number δ and intervals T_μ and T_ν – where T_ν is to the left of T_μ – such that for $v \in T_\nu, u \in T_\mu$

UNIMODALITY

(5.3.1) $$\begin{cases} \psi(v+\delta) - \psi(v) < \psi(u+\delta) - \psi(u) \\ \psi(v-\delta) - \psi(v) > \psi(u-\delta) - \psi(u). \end{cases}$$

We tentatively assume the contrary, that is, we suppose that for any pair of points v and u, $v < u$, there exists a sequence $\{\delta_n\}$ with $\delta_n > 0$, $\delta_n \to 0$, such that either

$$\psi(v+\delta_n) - \psi(v) \geq \psi(u+\delta_n) - \psi(u)$$

or $\quad \psi(v-\delta_n) - \psi(v) \leq \psi(u+\delta_n) - \psi(u).$

Suppose, for example, that the first of these two inequalities holds. Then for some u and v and for some sequence $\{\delta_{n_k}\}$

$$\frac{\psi(v+\delta_{n_k}) - \psi(v)}{\delta_{n_k}} \geq \frac{\psi(u+\delta_{n_k}) - \psi(u)}{\delta_{n_k}},$$

hence $\psi'(v) \geq \psi'(u)$.

Since this is true for all u and v it follows that $\psi(x)$ is a concave function. Therefore (5.3.1) holds when $\psi(x)$ is not concave, or by virtue of the definition of $\psi(x)$,

$$\frac{F'(v+\delta)}{F'(v)} < \frac{F'(u+\delta)}{F'(u)}$$

$$\frac{F'(v-\delta)}{F'(v)} > \frac{F'(u+\delta)}{F'(u)}$$

for all $v \in T_v$, $u \in T_u$. For such values of u and v we have

$$\left| \frac{F'(v+\delta)}{F'(v)} - \frac{F'(u+\delta)}{F'(u)} \right| \geq h > 0$$

$$\left| \frac{F'(v-\delta)}{F'(v)} - \frac{F'(u-\delta)}{F'(u)} \right| \geq h > 0.$$

Let u_0 and v_0 be the midpoints of T_u and T_v respectively. Let $T = [t_1, t_2]$ be an interval with centre at $(u_0 - v_0)/2$ such that $0 < t_1 < t_2$ and $x_0 + t \in T_u$, $x_0 - T \in T_v$ for $t \in T$. Denote by T^* the interval which is symmetric to T with respect to the origin. Intervals $[\alpha_1, \beta_1]$, $[\alpha_2, \beta_2]$ are chosen such that $\alpha_1 < \beta_1 < -t_2$, $t_2 < \alpha_2 < \beta_2$, and functions $\phi_1(t)$ and $\phi_2(t)$ are defined which are subject to certain conditions. For details of these we refer the reader to Ibragimov (1956). Let

$$R = T \cup T * \cup [\alpha_1, \beta_1] \cup [\alpha_2, \beta_2].$$

Define a function $g(t)$ in the following way:

$$g(t) = \begin{cases} -C_1(t-t_1)^2(t-t_2)^2 & \text{for } t \in T \\ C_2(t+t_1)^2(t+t_2)^2 & \text{for } t \in T^* \\ \phi_1(t) & \text{for } t \in [\alpha_1, \beta_1] \\ \phi_2(t) & \text{for } t \in [\alpha_2, \beta_2] \\ 0 & \text{for } t \in \bar{R} \end{cases}$$

Here C_1 and C_2 are constants defined in terms of t_1, t_2 and F'. It follows from the definition of $g(t)$ that

$$G(x) = \frac{1}{\displaystyle\int_{-\infty}^{\infty} du \int_{-\infty}^{u} g(t)\, dt} \int_{-\infty}^{x} du \int_{-\infty}^{u} g(t)\, dt$$

is a unimodal distribution function.

We consider the distribution function

$$H(x) = G(x) * F(x).$$

One can then show that $H''(x_0) = 0$ implies that for $\delta > 0$

$$H''(x_0 + \delta) > \frac{h}{2}\left[\int_{-\infty}^{\infty} du \int_{-\infty}^{u} g(t)\, dt\right]^{-1} > 0$$

and

$$H''(x_0 - \delta) < -\frac{h}{2}\left[\int_{-\infty}^{\infty} du \int_{-\infty}^{u} g(t)\, dt\right]^{-1}$$

Therefore $H(x)$ is not unimodal although $G(x)$ is unimodal. This means that the assumptions "$\psi(x)$ is not concave" and "$F(x)$ is strongly unimodal" are incompatible. Hence we have proved the following statement: If $F(x)$ is a strongly unimodal distribution function with a continuous second derivative on $E = (-\infty, \infty)$ then the logarithm of the density of $F(x)$ is concave on $(-\infty, \infty)$.

In order to remove assumption (A) two lemmas are used:

Lemma 5.3.1. The set of strongly unimodal distributions is closed under the operation of convolution and also under the operation of weak convergence.

Lemma 5.3.2. Let $\{F_n(x)\}$ be a sequence of strongly unimodal distribution functions. If

$$\lim_{n \to \infty} F_n(x) = F(x),$$

then there exists a subsequence such that

$$\lim_{k \to \infty} F'_{n_k}(x) = F'(x)$$

almost everywhere.

Lemma 5.3.1 is obvious; for the proof of Lemma 5.3.2, as well as for the removal of assumption (A), we refer the reader to Ibragimov (1956).

In the proof of the general case (that is in removing (A)) Theorem 5.3.1 is used.

We proceed to the discussion of the sufficiency of the condition of Theorem 5.3.2 and assume that $F(x)$ is strongly unimodal. Suppose that $G(x)$ is a sufficiently smooth distribution function [otherwise we consider $G(x) * \Phi(nx)$ instead of $G(x)$] with mode zero.

Let
$$H(x) = F(x) * G(x).$$

We have to show that $H''(x_0) = 0$ implies that $H''(x) \leq 0$ for $x > x_0$.

Let $\delta > 0$ and suppose that $H''(x_0) = 0$. Then it follows that

$$H''(x_0+\delta) = \int_0^\infty G''(x) F'(x_0-t) \frac{F'(x_0+\delta-t)}{F'(x_0-t)} dt$$

$$+ \int_0^\infty G''(-t) F'(x_0+t) \frac{F'(x_0+\delta+t)}{F'(x_0+t)} dt$$

$$= \frac{F'(x_0+\delta-\xi_1)}{F'(x_0-\xi_1)} \int_0^\infty G''(t) F'(x_0-t) dt$$

$$+ \frac{F'(x_0+\delta+\xi_2)}{F'(x_0+\xi_2)} \int_0^\infty G''(-t) F'(x_0+t) dt.$$

By an indirect proof one can show that for all ξ_1, ξ_2 and $\delta > 0$,

$$\frac{F'(x_0+\delta-\xi_1)}{F'(x_0-\xi_1)} \geq \frac{F'(x_0+\delta+\xi_2)}{F'(x_0+\xi_2)}.$$

Then one can conclude that $H''(x_0+\delta) \leq 0$ ($\delta > 0$). Finally if $E \neq (-\infty, \infty)$ it can be shown that there always exists a sequence $\{F_n(x)\}$ of strongly unimodal functions on $(-\infty, \infty)$ such that

$$\lim_{n \to \infty} F_n(x) = F(x).$$

5.4 Unimodality of stable and self-decomposable distributions

We mentioned in Section 5.2 that an incorrect proof of A. I. Lapin was originally used to prove the unimodality of all stable distributions. Subsequently I. A. Ibragimov and Yu. V. Linnik presented in their book [Russian original (1965), English translation (1971)] a proof for the unimodality of all stable distributions. But M. Kanter (1976) showed that this proof was also invalid.[*]

---

[*] The erroneous proof of Ibragimov and Linnik was reproduced in Lukacs (1970) and is omitted here.

M. Yamazato (1975) studied a class (the so-called L-class) of characteristic functions. This class contains all stable distributions and he succeeded in proving the unimodality of all distributions[*] belonging to this class. Yamazato's theorem therefore implies the unimodality of all stable distributions.

In recent years several papers have dealt with the question whether self-decomposable distributions were unimodal. Already A. Wintner (1956) had shown that every symmetric self-decomposable distribution is unimodal. Subsequently several authors tried to solve the general problem of the unimodality of self-decomposable distributions. The following partial result is due to S. J. Wolfe (1971a).

Theorem 5.4.1. *Suppose that $F(x)$ is a self-decomposable distribution and that the support[†] of its spectrum in the Lévy canonical representation is either the half line $(-\infty, 0)$ or the half line $(0, \infty)$; then $F(x)$ is unimodal.*

Using Theorem 5.2.1, Wolfe showed that every self-decomposable distribution function is the convolution of two unimodal distributions.

In proving Theorem 5.4.1 Wolfe (1971a) also obtained the following result which we shall use below. For its formulation we use the following notations:

Let $s_0 = 0 < s_1 = \ldots < s_k$, and let $\lambda_1, \lambda_2, \ldots, \lambda_k$ be positive constants. We write

$$\lambda = \sum_{i=1}^{k} \lambda_i$$

and

$$\lambda_0(x) = \begin{cases} \sum_{j=i}^{k} \lambda_j & \text{if } s_{i-1} \leq x < s_i \quad (i = 1, 2, \ldots, k) \\ 0 & \text{if } x \geq s_k. \end{cases}$$

Lemma 5.4.1. *Let γ be a constant and*

$$(5.4.1) \quad M_0(u) = \begin{cases} -\int_u^\infty x^{-1} \lambda_0(x) \, dx & \text{if } u > 0 \\ 0 & \text{if } u < 0 \end{cases}$$

and put

$$(5.4.2) \quad f_0(t) = \exp\{i\gamma t\} + \int_{-\infty}^{\infty} \left(e^{iu} - 1 - \frac{iut}{1+u^2} \right) dM_0(u).$$

Then $f_0(t)$ is the characteristic function of a self-decomposable distribution.

---

[*] Distributions of the L-class are also called self-decomposable distributions.

[†] A distribution function $G(x)$ is said to have an interval (α, β) as its support if $G(\alpha - 0) = 0$ while $G(\beta + 0) = 1$, that is, $\alpha = \text{lext } G$, $\beta = \text{rext } G$.

UNIMODALITY

Let $F_0(x)$ be the distribution function which corresponds to the characteristic function $f_0(t)$. Then

(a) $F_0(x)$ is absolutely continuous with density $q_0(x)$. The density $q_0(x)$ is continuous except at $x = 0$, and satisfies for $x > 0$ the differential equation

$$xq_0'(x) = \lambda q_0(x) - \lambda_1 q_0(x - s_1) - \ldots - \lambda_k q_0(x - s_k)$$

except at $x = 0, 1, \ldots, s_k$.

(b) If $\lambda \leqslant 1$ then $q_0'(x) \leqslant 0$, $q_0(x)$ non-increasing for $x > 0$.

(c) If $\lambda > 1$ then F_0 is unimodal with mode at $a > 0$ and $q_0(+0) = 0$.

M. Yamazato (1975) supplemented this by showing that

(d) $1 \leqslant |f_0| < \lambda \leqslant 2$, then $q_0(x)$ is concave on $(0, a]$.

We also shall need a condition which assures that the convolution of two unimodal distributions, with specified supports, is unimodal.

Theorem 5.4.2. Let $G(x)$ and $H(x)$ be unimodal distribution functions with modes a and b respectively. Suppose that the support of $G(x)$ is the half-line $(0, +\infty)$ while $H(x)$ has the support $(-\infty, 0)$. Assume that $G(x)$ and $H(x)$ are absolutely continuous with densities $p(x)$ and $q(x)$ respectively.

If $a > 0$ then we assume that $p(x)$ is log-concave on $(0, a]$, if $b < 0$ we assume that $q(x)$ is log-concave on $[b, 0)$; moreover we suppose that $q(b) = q(b+) \geqslant q(b-)$ and $a(-0) = 0$. Let

$$F = G * H = \int_{-\infty}^{\infty} G(x - y)\, dH(y), \quad \text{then } F(x) \text{ is unimodal.}$$

Proof. There are five cases:

(1) $a = b = 0$
(2) $a > 0 > b$ and $a + b \geqslant 0$
(3) $a > 0 > b$ and $a + b < 0$
(4) $a > 0 = b$
(5) $a = 0 > b$.

Let

$$r(x) = \int_{-\infty}^{\infty} p(x - y)\, q(y)\, dy \quad \text{be the density of } F(x).$$

Case (1). It follows that $r(x)$ is non-decreasing on $(-\infty, b)$ and non-increasing on (a, ∞), i.e. $F(x)$ is unimodal.

Case (2). Suppose that $p(x)$ and $q(x)$ are absolutely continuous on $(0, +\infty)$ respectively on $(-\infty, 0)$ with continuous derivatives $p'(x)$ and $q'(x)$ respectively. Then

$$(5.4.3) \quad r'(x) = \int_{-\infty}^{b} q'(y) p(x-y) \, dy + \int_{b}^{0 \wedge x} q'(y) p(x-y) \, dy.$$

Here $a \wedge x = \min(a, x)$; similarly $a \vee x = \max(a, x)$.

We first prove the following statements:

(A) If $r'(x_0) \geq 0$ for some $x_0 \in (b, a+b]$ then $r'(x_0 - \epsilon) \leq 0$ for any $\epsilon > 0$.
(B) If $r'(x_0) \leq 0$ for some $x_0 \in [a+b, a)$ then $r'(x_0 + \epsilon) \leq 0$ for any $\epsilon > 0$.

By assumption, $p'(x)$ is continuous on $(0, a]$. Let $\psi(x) = \log p(x)$; since $p(x)$ is log-concave, $\psi'(x)$ is non-increasing on $(0, a]$.

Therefore, if $0 < u < v < a - \epsilon$, one has

$$(5.4.4) \quad \psi(v+\epsilon) - \psi(u+\epsilon) = \int_{u+\epsilon}^{v+\epsilon} \psi'(x) \, dx \leq \int_{u}^{v} \psi'(x) \, dx = \psi(v) - \psi(u)$$

or

$$\psi(v+\epsilon) - \psi(v) \leq \psi(u+\epsilon) - \psi(u).$$

Hence

$$\frac{p(v+\epsilon)}{p(v)} \leq \frac{p(u+\epsilon)}{p(u)}.$$

We put

$$(5.4.5) \quad A_\epsilon(x) = \begin{cases} \dfrac{p(x+\epsilon)}{p(x)} & \text{if } p(x) > 0 \\ 0 & \text{if } p(x) = 0 \end{cases}$$

so that

$$(5.4.6) \quad A_\epsilon(v) \leq A_\epsilon(u).$$

By a limiting argument, (5.4.5) can be obtained in the case where $p(x)$ does not have a continuous derivative.

The function $p(x)$ is, by assumption, continuous and log-concave on $(0, a]$; therefore $A_\epsilon(x)$ is continuous and non-increasing for $0 \leq x \leq a - \epsilon$. Since $p(x)$ is non-decreasing on $[0, a]$ and non-increasing on $[a, \infty)$, one concludes that $A_\epsilon(x)$ is non-increasing on $[a-\epsilon, a]$ and that $A_\epsilon(x) \leq 1$ for $x > a$, but $A_\epsilon(x) \geq 1$ for $x \leq a - \epsilon$.

To prove (A), suppose that $r'(x_0) \geq 0$ for some $x_0 \in (b, a+b]$. It is already known that $r'(x_0 - \epsilon) \geq 0$ for $\epsilon \geq x_0 - b$. Thus it is only necessary to prove that $r'(x_0 - \epsilon) \geq 0$ when $\epsilon < x_0 - b$. If $\epsilon < x_0 - b$, then it follows from (5.4.2) and

(5.4.4) that

$$r'(x_0-\epsilon) = \int_{-\infty}^{b} \frac{p(x_0-y)}{A_\epsilon(x_0-\epsilon-y)} q'(y)\,dy$$

$$+ \int_{b}^{x_0-\epsilon} \frac{p(x_0-y)}{A_\epsilon(x_0-\epsilon-y)} q'(y)\,dy.$$

Since $q'(x)$ has no change of sign on $(-\infty, b)$ and on (b, ∞) we conclude from the properties of $A_\epsilon(x)$ that

(5.4.7) $\quad r'(x_0-\epsilon) = [A_\epsilon(x_0-\epsilon-\xi_1)]^{-1} \int_{-\infty}^{b} q'(y)\,p(x_0-y)\,dy$

$$+ [A_\epsilon(x_0-\epsilon-\xi_2)]^{-1} \int_{b}^{x_0-\epsilon} q'(y)\,p(x_0-y)\,dy,$$

where $x_0-b-\epsilon < a-\epsilon$ and $0 \leq x_0-\xi_2-\epsilon \leq x_0-b\ -\epsilon \leq x_0-\xi_1-\epsilon$. It follows from these inequalities that $A_\epsilon(x_0-\epsilon-\xi_2) \geq 1$. If $x_0-\epsilon-\xi_1 < a$ then $A_\epsilon(x_0-\epsilon-\xi_1) \leq A_\epsilon(x_0-\epsilon-\xi_2)$ since $A_\epsilon(x)$ is non-increasing on $(0, a)$. If $x_0-\epsilon-\xi_1 > a$ then $A_\epsilon(x_0-\epsilon-\xi_1) \leq 1$, so it is also true that $A_\epsilon(x_0-\epsilon-\xi_1) \leq A_\epsilon(x_0-\epsilon-\xi_2)$. It follows from (5.4.7) that

$$r'(x_0-\epsilon) \geq [A_\epsilon(x_0-\epsilon-\xi_2)]^{-1} \int_{-\infty}^{x_0-\epsilon} q'(y)\,p(x_0-y)\,dy$$

$$\geq [A_\epsilon(x_0-\epsilon-\xi_2)]^{-1} \int_{-\infty}^{x_0} q'(y)\,p(x_0-y)\,dy$$

$$= [A_\epsilon(x_0-\epsilon-\xi_2)]^{-1} r'(x_0) \geq 0.$$

This completes the proof of (A).

To prove (B), we apply (A) to $r(-x)$, the convolution of $p(-x)$ and $q(-x)$, and we use the reasoning which yielded (A). Alternatively one could employ the expression

$$r'(x) = \int_{0 \vee x}^{a} p'(y)\,q(x-y)\,dy + \int_{a}^{\infty} p'(y)\,q(x-y)\,dy$$

and use

$$B_\epsilon(x) = \begin{cases} q(x-\epsilon)/q(x) & \text{if } q(x) > 0 \\ 0 & \text{if } q(x) = 0 \end{cases}$$

and proceed as in the proof of statement (A).

We next show that $F(x)$ is unimodal. Let $M = \inf\{x : r'(x) < 0\}$; clearly $b \leq M \leq a$ and it follows from the continuity of $r'(x)$ that $r'(M) = 0$. We distinguish two cases:

Case (i). If $b \leq M \leq a+b$ it follows from (B) that $r'(x) \geq 0$ for $b \leq x \leq M$. Similarly, it follows from (B) that $r'(x) \leq 0$ if $x \in (M, a+b]$. To see this we note the following. If $r'(x_1) > 0$ for some $x_1 \in [M, a+b)$, then $r'(x_1) \geq 0$ in $[M, x_1]$, in contradiction to the definition of M. It follows that $r'(x) \leq 0$ in $[a+b, a)$.

Case (ii). If $a+b < M \leq a$ then the proof can be obtained in the same way as in Case (i).

In deriving (A) and (B) we made the assumption — not contained in the statement of Theorem 5.4.2 — that $p(x)$ and $q(x)$ are absolutely continuous. This assumption can be omitted. We note that $p(x)$ is log-concave on $(0, a]$, so that it is absolutely continuous on $(0, a]$. Then we can find a sequence $\{G_n(x)\}$ of absolutely continuous distribution functions with densities $p_n(x)$ and mode a such that

$$\lim_{n \to \infty} G_n(x) = G(x).$$

The densities $p_n(x)$ coincide with $p(x)$ on $(0, a]$ and are non-increasing step functions on $[a, \infty)$. For each $G_n(x)$ we can select a sequence $\{G_{nm}(x)\}$ of absolutely continuous unimodal distribution functions with mode a and densities $p_{nm}(x)$ such that

$$\lim_{m \to \infty} G_{nm}(x) = G_n(x).$$

The function $p_{nm}(x)$ is absolutely continuous on $(0, \infty)$ and coincides with $p(x)$ on $(0, a]$. We treat $H(x)$ in a similar way and obtain a sequence $\{H_{nm}(x)\}$ of unimodal distribution functions with mode b such that $H_{nm}(x) \to H(x)$ as $m \to \infty$, $n \to \infty$. The densities $q_{nm}(x)$ of $H_{nm}(x)$ are absolutely continuous and coincide with $q(x)$ on $[b, 0)$. Then $G_{nm} * H_{nm}$ is a unimodal distribution function which converges to $F(x) = G(x) * H(x)$ as $m \to \infty$, $n \to \infty$.

Since the weak limit of a sequence of unimodal distribution functions is unimodal [Theorem 4.5.4 of Lukacs (1970)] we conclude that $F(x)$ is unimodal.

Cases (iii), (iv) and (v) are treated in a similar way.

The problem of the unimodality of the L-class was solved by M. Yamazato (1975) who obtained the following result:

Theorem 5.4.3. All self-decomposable distributions are unimodal.

The proof of Theorem 5.4.3 makes strong use of Lemma 5.4.1 and of Theorem 5.4.2.

Let $G(x)$ be a distribution function which satisfies the conditions of Lemma 5.4.1 and which has the spectral function (5.4.1). Let $g(t)$ be the characteristic

UNIMODALITY

function of $G(x)$ so that

$$\log g(t) = \int_{+0}^{\infty} (e^{itu} - 1) u^{-1} \lambda(u) \, du.$$

The function $G(x)$ has all the properties required by Lemma 5.4.1.

Let $0 \geqslant v_1 > v_2 > \ldots > v_m$ and let $\mu_1, \ldots, \mu_m$ be positive constants and write

$$\mu = \sum_{j=1}^{m} \mu_j.$$

Define a function

$$\mu(x) = \begin{cases} \sum_{j=i}^{m} \mu_j & \text{if } v_i < x \leqslant v_{i-1} \\ 0 & \text{if } x \leqslant v_m. \end{cases}$$

Let

$$N(u) = \int_{-\infty}^{u} |x|^{-1} \mu(x) \, dx \quad (u < 0).$$

We construct a distribution function $H(x)$ with rext $H = 0$ and spectral function $N_0(u)$ such that

$$h(t) = \int_{-\infty}^{-0} (e^{itu} - 1) |u|^{-1} \mu(u) \, du$$

is the characteristic function of $H(x)$.

By reasoning similar to that which we used concerning $G(x)$, we see that $H(x)$ has all the properties required by Lemma 5.4.1. It follows from this lemma that $F = G * H$ is unimodal. Hence $F(x)$ is the self-decomposable distribution whose characteristic function $f(t)$ is given by

$$(5.4.8) \quad \log f(t) = \int_{-\infty}^{-0} \left(e^{itu} - 1 - \frac{itu}{1+u^2} \right) |u|^{-1} \mu(u) \, du$$

$$+ \int_{+0}^{+\infty} \left(e^{itu} - 1 - \frac{itu}{1+u^2} \right) u^{-1} \lambda(u) \, du.$$

Let a self-decomposable distribution $F(x)$ without normal component be given. By a suitable selection of the $\mu_n(u)$ and the $\lambda_n(u)$ one can arrange that

$$\lim_{n \to \infty} F_n(x) = \tilde{F}(x).$$

It follows that $\tilde{F}(x)$ is unimodal.

Let $F(x)$ be a general self-decomposable distribution and $f(t)$ be its characteristic function. Then one concludes from the fact that the normal distribution is strongly unimodal that $F(x)$ is unimodal so that Theorem 5.4.3 is established.

Theorem 5.4.3 is very significant for the following reason. The set of stable characteristic functions is a proper subset of the set of self-decomposable characteristic functions. Theorem 5.4.3 implies therefore the unimodality of all stable distributions. Thus this old problem has been finally settled.

K. Sato and M. Yamazato (1978) presented a deeper analysis of self-decomposable distributions. In this connection they introduced the concept of strictly unimodal distributions. A distribution function $F(x)$ is said to be strictly unimodal if there exists a point a such that $F(x)$ is absolutely continuous on $(-\infty, a) \cup (a, \infty)$ and if it has a density which is strictly increasing on (b_1, a) and strictly decreasing on (a, b_2). Here $b_1 = \text{lext } F$ while $b_2 = \text{rext } F$. The mode of a strictly unimodal distribution is unique. This is of course not necessarily true in the case where a distribution is unimodal but not strictly unimodal. Sato & Yamazato defined subsets of the family of self-decomposable distributions and studied these classes. They showed that with one exception all these subclasses are strictly unimodal. They also gave a necessary and sufficient condition for the existence of an unbounded density of a self-decomposable distribution and studied an integro-differential equation satisfied by self-decomposable distributions.

J. L. Wolfe (1971a) studied the analytical properties of characteristic functions of the L-class.

Theorem 5.4.4. Let $F(x)$ be a non-degenerate self-decomposable distribution function without a normal component and with spectral function $M(u)$. Let $\lambda(u) = uM'(u)$. A necessary and sufficient condition for $F(x)$ to have continuous derivatives of the first k orders is that $\lambda(+0) + |\lambda(-0)| > k$.

The sufficiency of the condition is a generalization of a result of V. M. Zolotarev (1963). In this paper Zolotarev also gives a necessary and sufficient condition for the boundedness of the density of a self-decomposable distribution.

Recently the concepts of self-decomposability and stability have been extended by F. W. Steutel and K. van Harn (1979) to discrete distributions on the non-negative integers.

6 Factorizations and infinite divisibility

In this chapter we discuss a number of somewhat loosely connected theorems. Some of these results are stated without proofs but with appropriate references.

6.1 Some properties of absolutely continuous or discrete infinitely divisible distributions

Theorem 6.1.1. Let k be any positive integer and let $F(x)$ be an infinitely divisible distribution function with characteristic function $f(t)$. The moment of order $(2k)$ of $F(x)$ exists if, and only if, the moment of order $(2k)$ of its spectral function exists.

The characteristic function $f(t)$ admits the Lévy-Khinchine representation

$$(6.1.1) \quad \log f(t) = iat + \int_{-\infty}^{\infty} \left(e^{itx} - 1 - \frac{itx}{1+x^2} \right) \frac{1+x^2}{x^2} \, d\theta(x).$$

To prove the sufficiency of the condition of Theorem 6.1.1 we assume

$$\int_{-\infty}^{\infty} x^{2k} \, d\theta(x) < \infty.$$

If this is the case then

$$\alpha(t) = \int_{-\infty}^{\infty} \left(e^{itx} - 1 - \frac{itx}{1+x^2} \right) \frac{1+x^2}{x^2} \, d\theta(x)$$

can be differentiated $(2k)$ times under the integral sign. This implies that $\dfrac{d^{2k}}{dt^{2k}} (\log f(t))$ exists and is finite. Hence $\int_{-\infty}^{\infty} x^{2k} \, dF(x)$ exists. To prove the necessity of the condition of the theorem we assume that

$$\int_{-\infty}^{\infty} x^{2k} \, dF(x) < \infty.$$

Therefore $f(t)$ also has a finite derivative of order $(2k)$ and this is also true for $\alpha(t)$. Then

$$\alpha''(t) = \lim_{h \to 0} \frac{\alpha(t+2h) - 2\alpha(t) + \alpha(t-2h)}{4h^2}.$$

Since $\alpha(0) = 0$ we have

$$\alpha''(0) = \lim_{h \to 0} \frac{\alpha(2h) + \alpha(-2h)}{4h^2}$$

$$= \lim_{h \to 0} \int_{-\infty}^{\infty} \frac{(e^{2ih} - 2 + e^{-2ih})}{4h^2} \frac{(1+x^2)}{x^2} d\theta(x)$$

$$= \lim_{h \to 0} \int_{-\infty}^{\infty} \left(\frac{e^{ih} + e^{-ih}}{2hx}\right)^2 (1+x^2) d\theta(x)$$

$$= -\lim_{h \to 0} \int_{-\infty}^{\infty} \left(\frac{\sin hx}{hx}\right)^2 (1+x^2) d\theta(x).$$

According to Fatou's lemma,

$$\int_{-\infty}^{\infty} (1+x^2) d\theta(x) \leq \lim_{h \to 0} \int_{-\infty}^{\infty} \left(\frac{\sin hx}{hx}\right)^2 (1+x^2) d\theta(x).$$

Therefore

$$\int_{-\infty}^{\infty} x^2 d\theta(x) < \infty.$$

This means that we can differentiate $\alpha(t)$ twice under the integral sign, so that

$$\alpha''(t) = -\int_{-\infty}^{\infty} (1+x^2) e^{itx} d\theta(x).$$

The last integral can be differentiated $(2k-2)$ times under the integral sign; hence

$$\int_{-\infty}^{\infty} x^{2k} d\theta(x) < \infty.$$

The unimodality of the spectral function of an infinitely divisible characteristic function was studied by C. Alf and T. O'Connor (1977). Their result is analogous to Theorem 5.1.1. They later derived a characterization of infinitely divisible characteristic functions with unimodal spectral function.

Conditions for the infinite divisibility of certain discrete distributions and of absolutely continuous distributions are also known. F. W. Steutel (1971) obtained the following results:

Theorem 6.1.2. Let $\{p_n\}$ be a probability distribution on the non-negative integers with $p_0 > 0$. The necessary and sufficient condition for the infinite *divisibility of the distribution* $\{p_n\}$ *is that*

$$np_n = \sum_{j=0}^{n-1} p_j q_{n-j-1}$$

where $q_j \geq 0$ $(j = 0, 1, \ldots)$ and $\sum_{j=1}^{\infty} \frac{1}{j} q_j < \infty$.

Steutel also obtained a similar result for distributions on $[0, \infty)$ and studied the zeros of infinitely divisible characteristic functions.

Theorem 6.1.3. *If* $\{p_n\}$ *is an infinitely divisible distribution on the non-negative integers and if* $p_0 > 0$, *then* $p_1 > 0$ *implies* $p_k > 0$ *for all* k.

Theorem 6.1.4. *Let* $p(x)$ *be a continuous density of an infinitely divisible distribution. Then* $p(x_0) = 0$ *implies that* $p(x) = 0$ *for all* $x \leq x_0$.

A function $g(x)$ defined on $(0, \infty)$ is said to be completely monotone if it has derivatives $g^{(n)}$ of all orders and if $(-1)^n g^{(n)}(x) \geq 0$.

Similarly, a sequence $\{x_n\}$ is said to be completely monotone if $(-1)^r \Delta^r x_k \geq 0$ for all r and all k.

The differences Δ^r are defined recursively, $\Delta^1 = \Delta$ and $\Delta^r = \Delta(\Delta^{r-1})$, $\Delta^0 x_n = x_n$. It can be shown by induction that

$$\Delta^r x_n = \sum_{k=0}^{r} \binom{r}{k} (-1)^{r+k} x_{n+k}.$$

The following theorem is due to F. W. Steutel (1969).

Theorem 6.1.5. *All completely monotone densities are infinitely divisible. All completely monotone lattice distributions are infinitely divisible.*

An extensive discussion of infinitely divisible distributions defined on the set of non-negative integers can be found in K. van Harn's (1978) monograph.

Two distribution functions are said to be equivalent if they are absolutely continuous with respect to each other.

The equivalence of infinitely divisible distributions was studied by W. N. Hudson and H. G. Tucker (1975). They also studied the equivalence of the Lévy spectral measure to the Lebesgue measure.

It is often difficult to decide whether a given density function is infinitely divisible or not and to devise methods to answer this question. A number of authors have investigated problems of this kind.

The infinite divisibility of Student's t-distribution was studied by E. Grosswald (1976a, b), M. E. H. Ismail (1977), and M. E. H. Ismail & D. Kelker (1976).

O. Thorin (1977b) showed that the lognormal distribution[*] is infinitely divisible and also [Thorin (1977a)] that the Pareto distribution[†] is infinitely divisible. T. Lewis (1976) showed that the von Mises distribution[§] is infinitely divisible for certain values of the parameter (namely if k is sufficiently small). O. Barndorff-Nielsen & C. Halgreen (1977) studied the infinite divisibility of the hyperbolic and of the generalized inverse Gaussian distributions. S. G. Maloshevski (1972) considered densities of the form

$$p(x) = C \exp[\beta x - A e^{\alpha x}]$$

with $A > 0$ and $\alpha\beta > 0$ and showed that they are infinitely divisible. L. Bondesson (1979a, b) generalized Thorin's work by studying generalized Gamma convolutions already introduced by O. Thorin (1977a).

The following fact is well known and easily proved. If F is a symmetric distribution then $F * F$ is also symmetric. R. G. Staudte and M. N. Tata (1970) studied the problem whether the converse is true, that is, under what conditions it is possible to conclude from the assumption that $F * F$ is symmetric that F is symmetric. They constructed examples of non-symmetric distributions F such that $F * F$ is symmetric. This established the need for some supplementary conditions, and they obtained the following result.

Suppose that the characteristic function $f(t)$ of a distribution F is either infinitely divisible or is an analytic characteristic function, then one can conclude from the assumption that $F * F$ is symmetric that F is also symmetric.

The definition of an analytic characteristic function is given in the introductory paragraph of Chapter 7.

6.2 Tail behaviour of infinitely divisible distributions

Let $F(x)$ be a distribution function. We write $T(x) = 1 - F(x) - F(-x)$ and call $T(x)$ the tail of the distribution function $F(x)$.

In the following we shall present some results concerning the behaviour of the tail of an infinitely divisible distribution. More precisely, we shall study the

---

[*] The lognormal distribution is defined by its density

$$p(x) = \frac{1}{\sigma(x-a)\sqrt{2\pi}} \exp\left\{-\frac{[\log(x-a)-u]^2}{2\sigma^2}\right\} \text{ if } x > a;$$

$p(x) = 0$ if $x \leq a$.

[†] The Pareto distribution has the density $p(x) = 1 - \left(\frac{x}{\alpha}\right)^{-h}$ if $x \geq \alpha$; $p(x) = 0$ if $x < \alpha$. It is a special case of beta distributions.

[§] The density of the von Mises distribution is

$$p(x) = \frac{1}{2\pi I_0(k)} \exp(k \cos y) \text{ if } -\pi < x < +\pi, k > 0.$$

I_0 is a Bessel function. In a subsequent paper, Lewis (1976) showed that the von Mises distribution is infinitely divisible for $k < 0 \cdot 16$.

FACTORIZATIONS AND INFINITE DIVISIBILITY 69

question how fast the tail of an infinitely divisible distribution can approach zero.

Theorem 6.2.1. Suppose that F(x) is infinitely divisible and that there exist positive constants a and α such that

$$T(x) = O[\exp(-ax^{1+\alpha})] \quad \text{as } x \to \infty.$$

Then

(i) *F(x) is a degenerate distribution if* $\alpha > 1$,
(ii) *F(x) is a normal distribution if* $0 < \alpha \leq 1$.

For the proof of the theorem we need the following lemma from the theory of entire functions [R. P. Boas (1954), p. 3].

Lemma 6.2.1. Suppose that f(z) is an entire function and that

$$\text{Re } f(z) \leq A(\epsilon) r^{\rho + \epsilon}$$

for any $\epsilon > 0$ and arbitrarily large r; then f(z) is a polynomial of degree at most equal to ρ.

We proceed to the proof of the theorem. We assumed that $T(x) = O[\exp(-ax^{1+\alpha})]$ as $x \to \infty$; it follows from Lemma 7.2.2 of Lukacs (1970) that the characteristic function $f(t)$ of $F(x)$ is an entire function of order $\rho \leq 1 + \alpha^{-1}$. Since $f(t)$ belongs to an infinitely divisible distribution it has no zeros, so that $f(z) = \exp[g(z)]$ where $g(z)$ is the principal value of $\log f(z)$ and $g(0) = 0$. Then

$$\max_{|z|=r} \text{Re } g(z) = \max_{|z|=r} \log |f(z)| = \log \max_{|z|=r} |f(z)| = r^{\rho + \epsilon}$$

for every sufficiently large r. According to Lemma 6.2.1 $g(z)$ is a polynomial of degree ρ. Theorem 1.4.4 (Marcinkiewicz' theorem) implies that $\rho \leq 2$ and the statements of Theorem 6.2.1 follow immediately.

If $F(x)$ is a finite distribution, that is, if $T(x) = 0$ for sufficiently large $|x|$, then condition (i) is satisfied for any α and we obtain

Corollary 1 to Theorem 6.2.1. A non-degenerate finite distribution cannot be infinitely divisible.

Corollary 2 to Theorem 6.2.1. The only infinitely divisible entire characteristic functions of finite order are the normal and the degenerate distribution.

A. Ruegg (1970) proved a generalization of Theorem 6.2.1 and obtained the following result:

Theorem 6.2.2. Let f(z) be an entire characteristic function which has no zeros and which does not belong to a degenerate distribution. Suppose that there exist constants $\alpha > 0$ *and* $\delta > 1$ *such that*

$$T(x) = O\{\exp[-\alpha x(\log x)^\delta]\} \quad \text{as } x \to \infty.$$

Then $f(t)$ *belongs to a normal distribution.*

The asymptotic behaviour of the tail $T(x)$ of a distribution function $F(x)$ was also studied by F. W. Steutel (1974) as well as by R. A. Horn (1972).

A generalization of Theorem 6.2.2 has been given by Horn (1972).

Theorem 6.2.3. Let $F(x)$ *be an infinitely divisible distribution function such that*

$$T(x) = O\{\exp[-xM(x)]\}$$

as $x \to \infty$, *where* $M(x)$ *is a non-negative measurable function.*

(i) $F(x)$ *is normal (possible degenerate) if, and only if,* $M(x)/\log x \to \infty$ *as* $x \to \infty$ *and if* $M(x)$ *is continuous and strictly monotone-increasing for sufficiently large* x.

(ii) $F(x)$ *is degenerate if and only if condition* (i) *is satisfied and if* $M(x)/x \to \infty$ *as* $x \to \infty$.

Theorem 6.2.2 is obtained from Theorem 6.2.3 by putting $M(x) = ax^\alpha$ with $a, \alpha > 0$ respectively, $M(x) = a(\log x)^\alpha$ if $a > 0$, $\alpha > 1$.

The proof of Theorem 6.2.3 uses properties of analytic characteristic functions. Since these will only be discussed in Chapter 7 we stated the theorem without proof.

6.3 Decomposition of symmetric stable distributions

Some symmetric stable distributions admit interesting decompositions. These are based on the extensions of Pólya's theorem [Theorem 1.2.2] discussed in Section 1.2. We give two examples.

Theorem 6.3.1. Let $g(t)$ *be a characteristic function without zeros. Assume further that* $f(t) = \log|g(t)|$ *has bounded first and second derivatives for* $t > 0$ *and that* $f(t) = O(t)$ *as* $t \to \infty$. *Then* $g(t)$ *is a factor of the Cauchy distribution* $\exp(-\lambda|t|)$, *provided that the parameter* λ *is sufficiently large.*

Since $g(t)$ has no zeros, $\log|g(t)|$ is defined for all t and we put

$$h(t) = -2f(t) = \log|g(t)|^{-2}.$$

Then

$$e^{h(t)} = [g(t)\,\overline{g(t)}]^{-1}$$

and

$$\exp(-\lambda|t|) = g(t)\,\overline{g(t)} \exp[-\lambda|t| + h(t)].$$

FACTORIZATIONS AND INFINITE DIVISIBILITY

The function $h(t)$ satisfies the conditions of Corollary 2 to Theorem 1.2.2, so that $\exp[-\lambda|t| + h(t)]$ is a characteristic function, provided λ is sufficiently large. Hence $g(t)$ is a factor of the symmetric Cauchy distribution.

Let $g(t) = \exp[\mu(e^{it} - 1)]$ be the characteristic function of the Poisson distribution. The conditions of Theorem 6.3.1 are then satisfied and we obtain the following statement.

Corollary to Theorem 6.3.1. *The Cauchy distribution has a Poissonian factor if the parameter λ is sufficiently large.*

Theorem 6.3.2. *Let $g(t)$ be a characteristic function without zeros. Suppose that $f(t) = \log|g(t)|$ has bounded first and second derivatives for $t > 0$. Assume further that $0 < \alpha < 1$ and that $f^{(k)}(t) = O(t^{\alpha-k})$ as $t \to \infty$ for $k = 0, 1, 2$. Then $g(t)$ is a factor of the symmetric stable distribution with characteristic function $\exp[-\lambda|t|^\alpha]$, provided the parameter λ is sufficiently large.*

We put

$$h(t) = -2f(e^t)e^{-\alpha t} = e^{-\alpha t}\log|g(e^t)|^{-2}.$$

Then

$$\exp[-\lambda + k(\log|t|)^\alpha] = \exp(-\lambda|t|^\alpha)[g(t)]^{-2}.$$

Therefore

$$\exp(-\lambda|t|^\alpha) = g(t)\,\overline{g(t)}\,\exp\{[-\lambda + h(\log|t|)]|t|^\alpha\}.$$

The theorem then follows easily from Corollary 3 to Theorem 1.2.2.

R. Shimizu (1972) obtained additional results of this type and also a generalization of Corollary 3 to Theorem 1.2.2.

6.4 Certain indecomposable distributions

We consider the frequency function of the beta distribution

$$(6.4.1)\quad p_{\alpha\beta}(x) = \begin{cases} \dfrac{\Gamma(\alpha+\beta)}{\Gamma(\alpha)\,\Gamma(\beta)}(1-x)^{\alpha-1}x^{\beta-1} & \text{if } 0 < x < 1 \\ 0 & \text{otherwise} \end{cases}$$

where $\alpha > 0, \beta > 0$.

Theorem 6.4.1. *The beta distribution is indecomposable if $\alpha + \beta < 2$.*

For the proof of this theorem we need the following lemma.

Lemma 6.4.1. *Let $F(x)$ be a decomposable finite distribution with* lext $F = m$, rext $F = M$. *Then there exist two numbers k_1 and $k_2, m < k_j < M$ ($j = 1, 2$) such* that for all $\epsilon > 0$ the inequality

$$[F(k_1+2\epsilon) - F(k_1-2\epsilon)][F(k_2+2\epsilon) - F(k_2-2\epsilon)]$$
$$\geq [1 - F(M-\epsilon)] F(m+\epsilon).$$

holds.

Let F_1 and F_2 be two non-singular factors of F such that

(6.4.2) $\quad F = F_1 * F_2$

and denote by m_1 and m_2 (resp. M_1 and M_2) the left (resp. the right) extremities of F_1 and F_2. Then $m_j < M_j$ ($j = 1, 2$), $M = M_1 + M_2$, $m = m_1 + m_2$. Equation (6.4.2) implies the following inequalities:

(6.4.3a) $\quad F(\xi + \eta + 2\epsilon) - F(\xi + \eta - 2\epsilon)$
$$\geq [F_1(\xi + \epsilon) - F_1(\xi - \epsilon)][F_2(\eta + \epsilon) - F_2(\eta - \epsilon)].$$
(6.4.3b) $\quad F(m + \epsilon) \leq F_1(m_1 + \epsilon) F_2(m_2 + \epsilon).$
(6.4.3c) $\quad 1 - F(M - \epsilon) \leq [1 - F_1(M_1 - \epsilon)][1 - F_2(M_2 - \epsilon)].$

Here ξ and η are arbitrary, $\epsilon > 0$. We substitute $\xi = m_1$, $\eta = M_2$ into (6.4.3a) and get

$$F(M_1 + m_2 + 2\epsilon) - F(M_1 + m_2 - 2\epsilon) \geq F_1(m_1 + \epsilon)[1 - F_2(M_2 - \epsilon)].$$

Similarly, substituting $\xi = M_1$, $\eta = m_2$ into (6.4.3a) we obtain

$$F(M_1 + m_2 + 2\epsilon) - F(M_1 + m_2 - 2\epsilon) \geq [1 - F_1(M_1 - \epsilon)] F_2(m_2 + \epsilon).$$

We multiply these two inequalities. In view of (6.4.3b) and (6.4.3c) we get

$$[F(m_1 + M_2 + 2\epsilon) - F(m_1 + M_2 - 2\epsilon)][F(M_1 + m_2 + 2\epsilon) - F(M_1 + m_2 - 2\epsilon)]$$
$$\geq [1 - F(M - \epsilon)] F(m + \epsilon).$$

If we put here $k_1 = m_1 + M_2$ and $k_2 = M_1 + m_2$ we obtain the statement of the lemma.

To prove the theorem we apply the lemma to the distribution function defined by the frequency function (6.4.1) and see that

$$16 \left\{ \frac{1}{4\epsilon} \int_{k_1 - 2\epsilon}^{k_1 + 2\epsilon} p_{\alpha\beta}(x) \, dx \right\} \left\{ \frac{1}{4\epsilon} \int_{k_2 - 2\epsilon}^{k_2 + 2\epsilon} p_{\alpha\beta}(x) \, dx \right\}$$
$$\geq \left\{ \frac{1}{\epsilon} \int_0^{\epsilon} p_{\alpha\beta}(x) \, dx \right\} \left\{ \frac{1}{\epsilon} \int_{1-\epsilon}^{1} p_{\alpha\beta}(x) \, dx \right\}.$$

Let $\epsilon \to 0$. The left-hand side has a finite limit, while the right-hand side tends to ∞, provided that $\alpha + \beta < 2$. This contradiction proves that the beta distribution with $\alpha + \beta < 2$ is indecomposable.

FACTORIZATIONS AND INFINITE DIVISIBILITY 73

If $\alpha = \beta = \frac{1}{2}$ then (6.4.1) becomes

$$p_{1/2\,1/2}(x) = \frac{1}{\pi}(1-x)^{-1/2}x^{-1/2} \quad \text{if } 0 < x < 1,$$

$p(x) = 0$ otherwise.

This is the frequency function of the arcsine distribution, so that we obtain the following corollary:

Corollary to Theorem 6.4.1. The arcsine distribution is indecomposable.

The question whether there exist indecomposable distributions whose spectrum is a given set has been studied by several authors. Lévy (1952) showed that for any closed set A there exists an indecomposable, absolutely continuous distribution whose spectrum is the set A. Dugué (1951b) constructed an absolutely continuous indecomposable distribution whose spectrum is the real line. We have discussed this example above.

Theorem 6.2.3 and its corollary, as well as Theorem 6.2.4 (all these in Lukacs, 1970), show that an infinitely divisible distribution (characteristic) function can have indecomposable factors.

Let F be a distribution which has indecomposable factors and denote the set of all indecomposable factors of F by $N(F)$. A description of the set $N(F)$ is known only for very few distributions. Let S be a closed set and denote by $\mathfrak{Z}_S$ the set of all distributions whose spectrum is S; then $N_S(F) = \mathfrak{Z}_S \cap N(F)$ is the set of all indecomposable factors of F whose spectrum is S.

A. I. Ilinskii (1974, 1977) has studied the set $N_S(F)$ for certain infinitely divisible distributions. We mention here a few of his results.

Let $G_{a\lambda}(x)$ be the geometric distribution, defined on the set $\{\lambda_k\}$ ($k = 0, 1, 2, \ldots; 0 < \lambda < \infty; 0 < a < 1$) of equidistant points, so that

$$G_{a\lambda}(x) = \sum_{k=0}^{\infty} (1-a)\,a^k \in (x - \lambda_k).$$

Let $H_a(x)$ be an exponential distribution with frequency function

$$p_a(x) = \begin{cases} a\,e^{-ax} & \text{for } x \geq 0 \\ 0 & \text{for } x < 0. \end{cases}$$

The distributions $G_{a\lambda}(x)$ and $H_a(x)$ are both infinitely divisible; $G_{a\lambda}(x)$ is a discrete distribution while $H_a(x)$ is absolutely continuous.

Theorem 6.4.2. Let S be an arbitrary subset of $\{\lambda_k\}_{k=0}^{\infty}$ which contains at least two points; then the set $N_S(G_{a\lambda})$ is not empty.

Theorem 6.4.3. Let $S \subset [0, \infty)$ be a closed set which contains at least two points and suppose that $0 \in S$. Then $N_S(H_a)$ is not empty.

Ilinskii also studied the set $N_S(F)$ of indecomposable factors for some other infinitely divisible distributions such as the distributions with characteristic functions $(1 + t^2)^{-1}$, $(\cosh t)^{-1}$, $\exp(-a|t|)$ with $0 < a \leq 1$, $(1 + |t|)^{-\gamma}$ with $\gamma > 2$. If S is a closed set which contains at least two points then the set $N_S(F)$ is not empty for these distributions.

S. Mase (1975) has considered infinitely divisible characteristic functions of the form

$$(6.4.4) \quad f(t) = \exp\left[iat + \fint_{-\infty}^{\infty} \left(e^{itu} - 1 - \frac{itu}{1+u^2}\right) \alpha(u)\, du \right].$$

Here $\fint_{-\infty}^{\infty}$ stands for $\int_{-\infty}^{-0} + \int_{+0}^{\infty}$ and $\alpha(u)$ is non-negative and

$$(6.4.4a) \quad \int_{|u| \geq 1} \alpha(u)\, du < \infty, \quad \int_{|u| \leq 1} u^2 \alpha(u)\, du < \infty.$$

Formula (6.4.4) is the Lévy canonical representation of an infinitely divisible law with $\sigma^2 = 0$ for which both $M(u)$ and $N(u)$ are absolutely continuous. In the paper quoted above, Mase gave the following generalizations of the preceding theorems.

Theorem 6.4.4. *Let $f(t)$ be a characteristic function defined by (6.4.4). Suppose that the set of all positive, as well as the set of all negative x on which $\alpha(x) > 0$ has a positive Lebesgue measure. Then $f(t)$ has an indecomposable factor.*

Theorem 6.4.5. *Let $f(t)$ be a non-degenerate characteristic function of the form (6.4.4). Suppose that $\alpha(x) = 0$ almost everywhere for $x < 0$ and that there is no positive number c such that $\alpha(x) = 0$ almost everywhere outside $[c, 2c]$. Then $f(t)$ has an indecomposable factor. This conclusion also holds if $\alpha(x) = 0$ almost everywhere for $x > 0$ and if there is no $c > 0$ such that $\alpha(x) = 0$ outside $[-2c, -c]$.*

The proof of Theorems 6.4.4 and 6.4.5 is somewhat similar to the proof of Theorem 6.2.3 in Lukacs (1970).

L. S. Kudina (1972) studied the spectra of indecomposable laws and obtained the following results.

Theorem 6.4.6. *Suppose that an arbitrary closed set A is given. Then there exists an indecomposable distribution function F such that $S(F) = A$.*

FACTORIZATIONS AND INFINITE DIVISIBILITY

Theorem 6.4.7. *Let A be a closed unbounded set and let* $1 < \rho \leqslant \infty$ *and* $0 \leqslant \tau \leqslant \infty$ *be two given numbers. Then there exists an indecomposable distribution function F for which* $S(F) = A$ *and whose characteristic function is an entire function of order ρ and type τ.*

In a subsequent paper Kudina (1973)[*] showed that the set of indecomposable distributions is dense, in the sense of weak convergence, in the set of all distributions whose spectrum is the fixed, non-empty closed set A.

---

[*] Actually, the paper deals with the multivariate case. If A is bounded then one also assumes that it is non-denumerable.

7 Analytic characteristic functions

In Section 1.4 we introduced analytic characteristic functions and we have discussed them in detail in Lukacs (1970). In this chapter we present a number of results which have been obtained since the publication of the monograph mentioned above.

7.1 Analytic continuation of analytic characteristic functions

Analytic characteristic functions can often be continued analytically beyond their strip of regularity. For example,

$$(7.1.1) \quad f(t) = \left[\left(1 - \frac{it}{a}\right)\left(1 + \frac{it}{b}\right)\right]^{-1} \quad (a > 0, b > 0)$$

is an analytic characteristic function with the strip $-a < \operatorname{Im}(z) < b$. It is clear that this rational function can be continued into the entire complex z-plane, although its integral representation holds only in its strip of regularity.

This situation leads to the question whether an analytic characteristic function can have a natural domain of analyticity, that is, a domain beyond which it cannot be continued analytically. The accessible points of the boundary of this domain are then singular points.

Theorem 7.1.1. Let G be a domain in the complex z-plane which contains the strip $-\alpha < \operatorname{Im}(z) < \beta$ and which is symmetric with respect to the imaginary axis. Suppose further that the points $i\beta$ and $-i\alpha$ belong to the boundary ∂G of G. Then there exists an analytic characteristic function whose natural domain of analyticity is G.

We consider first the case where the strip of regularity is the half-plane $\operatorname{Im}(z) < \beta$ [i.e. where $\alpha = -\infty$].

Theorem 7.1.2. Let G be a domain in the complex z-plane which is symmetric with respect to the imaginary axis and whose boundary ∂G is contained in the half plane $\operatorname{Im}(z) \geq \beta > 0$. Suppose that the point $i\beta$ belongs to the boundary of G. Then there exists an analytic characteristic function whose natural domain of analyticity is G.

For the proof of Theorem 7.1.2 we need several lemmas.

ANALYTIC CHARACTERISTIC FUNCTIONS

Lemma 7.1.1. *The function*

(7.1.2) $$f(t) = \left[\left(1 + \frac{it}{a}\right)\left(1 + \frac{it}{c+id}\right)\left(1 + \frac{it}{c-id}\right)\right]^{-1},$$

where a, c and d are real numbers, is a characteristic function if $0 < a \leqslant c$.

The lemma is proved by expanding $f(t)$ into partial fractions and applying the inversion formula.

Lemma 7.1.2. *It is always possible to find a sequence $\{a_k\}$ of positive numbers such that*

$$\sum_{k=1}^{\infty} a_k = 1 \quad \text{while} \quad a_k \geqslant 2 \sum_{k=n+1}^{\infty} a_k.$$

An example of such a sequence is $a_k = (c-1) c^{-k}$ with $c \geqslant 3$.

Lemma 7.1.3. *Let G be an arbitrary domain; then there exists a set of points which is dense in the boundary ∂G of G.*

We consider the set of all points of G with rational coordinates; they can be arranged in a sequence $\{\alpha_n\}$. For each point α_n let β_n be the point of the boundary of G which is nearest to the origin and for which $\arg(\alpha_n - \beta_n)$ has the smallest possible value. In general $\{\beta_n\}$ contains repetitions. We omit these and obtain a sequence $\{b_n\}$ which is dense in ∂G.

We first prove Theorem 7.1.2. The symmetry of G and of ∂G follows immediately from the Hermitian property of characteristic functions. We have assumed that ∂G is contained in the half-plane $\text{Im}(z) \geqslant \beta > 0$. Then $i\beta$ is necessarily a singular point (possibly an isolated singularity) of the analytic characteristic function which we wish to construct.

Let $\{b_n\}$ be the dense sequence of points of ∂G whose existence is established by Lemma 7.1.3. We suppose $b_n \neq i\beta$ and introduce the functions

(7.1.2a) $f_0(z) = (1 - z/i\beta)^{-1}$

and

(7.1.2b) $$f_n(z) = \left[\left(1 - \frac{z}{i\beta}\right)\left(1 - \frac{z}{b_n}\right)\left(1 - \frac{z}{\bar{b}_n}\right)\right]^{-1}.$$

Since $\text{Im}(b_n) \geqslant \beta$, we conclude from Lemma 7.1.1 that $f_n(z)$ is an analytic characteristic function.

We expand $f_n(z)$ into partial fractions and obtain

(7.1.3) $$f_n(z) = \frac{A_n}{z - i\alpha} + \frac{B_n}{z - b_n} + \frac{C_n}{z - \bar{b}_n}$$

where

(7.1.4)
$$\begin{cases} A_n = \dfrac{i\beta b_n \bar{b}_n}{(i\beta - b_n)(i\beta + \bar{b}_n)} \\ \\ B_n = \dfrac{i\beta b_n \bar{b}_n}{(b_n - i\beta)(b_n + \bar{b}_n)} = -\bar{C}_n. \end{cases}$$

We select a sequence $\{a_n\}$ which satisfies the conditions of Lemma 7.1.2. Then

(7.1.5) $\quad f(z) = \sum\limits_{n=0}^{\infty} a_n f_n(z)$

is an analytical characteristic function whose strip of regularity is the half-plane $\mathrm{Im}(z) < \beta$.

We next show that $f(z)$ is analytic in the domain G. We do this by proving that the series (7.1.5) is uniformly convergent in every domain $G_{\epsilon,R}$ of the form

$$G_{\epsilon,R} = \{z : z \in G, |z| < R, d(z, \partial G) > \epsilon\}.$$

Here $d(z, \partial G)$ is the distance of the point z from the boundary ∂G, while ϵ and R are arbitrary positive constants.

It follows from the relations (7.1.4) that

$$\frac{A_n}{z - i\beta} = O(1)$$

$$\frac{B_n}{z - b_n} = o(1)$$

$$\frac{C_n}{z - \bar{b}_n} = o(1)$$

as $|b_n| \to \infty$. Therefore it follows from (7.1.4) that

$$f_n(z) = O(1) \qquad \text{as } |b_n| \to \infty.$$

This means that there exist constants K and M such that

(7.1.6) $\quad |f_n(z)| < K \qquad \text{if } |b_n| > M.$

Let n be such that $|b_n| \leq M$. It follows from (7.1.2b) that

(7.1.7) $\quad |f_n(z)| = \left| \dfrac{\beta b_n \bar{b}_n}{(z - i\beta)(z - b_n)(z + \bar{b}_n)} \right| < \dfrac{\beta M^2}{\epsilon^3}.$

The uniform convergence of the series (7.1.5) in $G_{\epsilon,R}$ follows from (7.1.6) and (7.1.7).

ANALYTIC CHARACTERISTIC FUNCTIONS

We show next that each b_n is a singular point of $f(z)$. In the following we consider values of z for which

(7.1.8) $\quad |z - b_n| < |z - b_k|$

for all $k \neq n$. The construction of the sequence $\{b_k\}$ assures the existence of such z. It follows from (7.1.2b) and (7.1.8) that

$$|f_n(z)| \leqslant \left| \frac{A_k}{z - i\beta} + \frac{C_k}{z + b_n} \right| + \left| \frac{B_k}{z - b_n} \right|$$

so that

(7.1.9) $\quad \left| \frac{f_k(z)}{B_n/(z - b_n)} \right| \leqslant \left| \frac{\dfrac{A_k}{z - i\beta} + \dfrac{C_k}{(z + b_k)}}{B_n/(z - b_n)} \right| + \left| \frac{B_k}{B_n} \right|$

After some elementary but somewhat tedious computations one sees that it is possible to find positive constants δ and η_1 sufficiently small for one to conclude from (7.1.9) that

(7.1.10) $\quad \left| \dfrac{f_k(z)}{B_n/(z - b_n)} \right| < \dfrac{3}{2},$

provided that $0 < |z - b_n| < \eta_1$ and $|b_k - b_n| < \delta$.

It is also possible to select an $\eta_2 > 0$ sufficiently small that

(7.1.11) $\quad \left| \dfrac{f_n(z)}{B_n/(z - b_n)} \right| > \dfrac{2}{3}$

for $|z - b_n| < y_2$. It follows from (7.1.10) and (7.1.11) that

(7.1.12) $\quad \left| \dfrac{f_k(z)}{f_n(z)} \right| < 1,$

provided that $|z - b_n| < \eta \leqslant \min(\eta_1, \eta_2)$ and $|b_k - b_n| < \delta$. Inequality (7.1.12) remains valid if $|b_k - b_n| \geqslant \delta$ and if η is chosen sufficiently small.

We see from (7.1.5) that

(7.1.13) $\quad \dfrac{f(z)}{f_n(z)} = \sum_{k=0}^{n-1} a_k \dfrac{f_k(z)}{f_n(z)} + a_n + \sum_{k=n+1}^{\infty} a_k \dfrac{f_k(z)}{f_n(z)}.$

The first sum in (7.1.13) tends to zero as $z \to b_n$, while it follows from Lemma 7.1.2 and formula (7.1.13) that

$$\sum_{k=n+1}^{\infty} a_k \dfrac{f_k(z)}{f_n(z)} < \dfrac{a_n}{2}.$$

Therefore it follows from (7.1.12) that $|f(z)/f_n(z)| > a_n/4$ if $z \to b_n$. Therefore b_n is a singular point of $f(z)$.

We still have to consider the point $i\beta$. If $i\beta$ is not an isolated point of ∂G then it is the limit of singular points and therefore also a singular point of $f(z)$. Suppose next that $i\beta$ is an isolated point of ∂G. It is easily seen that $A_n = -imn$, where

$$m_n = \frac{\beta |b_n|^2}{|b_n - i\beta|^2} > 0$$

and we see from (7.1.4) and (7.1.5) that $i\beta$ is a singular point of $f(z)$. This completes the proof of Theorem 7.1.2.

A similar result can be obtained in the case where the boundary of the domain G is contained in the half-plane $\mathrm{Im}(z) \leq \alpha < 0$. Combining this with Theorem 7.1.2 we obtain Theorem 7.1.1.

7.2 Entire characteristic functions satisfying certain conditions

The conditions discussed in this section involve mostly the tail behaviour of the corresponding distribution function.

Theorem 7.2.1. *The distribution function $F(x)$ has an entire characteristic function $f(t)$ such that*

$$\limsup_{r \to \infty} \frac{\log \log \log M(r, f)}{\log r} = \alpha$$

if and only if

(i) $T(x) > 0$ for all $x > 0$

and

(ii) $\displaystyle\liminf_{x \to \infty} \frac{\log\left[\dfrac{1}{x} \log \dfrac{1}{T(x)}\right]}{\log \log x} = 1/\alpha.$

Proof. The necessity of (i) follows from Theorem 1.4.5. The necessity of (ii) follows from the assumption of the theorem and the easily established inequality $M(r; f) \geq \frac{1}{2} T(x) e^{rx}$. First we see that

$$T(x) \leq 2 e^{-rx} M(r, f) \leq 2 \exp[\exp(r^{\alpha+\epsilon}) - rx]$$

for all sufficiently large r (say $r \geq R$) and all $x \geq R$. Let $x \geq \exp(R^{\alpha+\epsilon})$ and $r = [\log x]^{1/(\alpha+\epsilon)}$; then $r \geq R$ and one sees that

$$T(x) \leq 2 \exp\{x - x[\log x]^{1/(\alpha+\epsilon)}\}.$$

ANALYTIC CHARACTERISTIC FUNCTIONS

It follows that
$$\liminf_{x\to\infty} \frac{\log\left[\frac{1}{x}\log\frac{1}{T(x)}\right]}{\log\log x} \geq 1/(\alpha+\epsilon).$$

Since $\epsilon > 0$ is arbitrary one concludes that

(7.2.1) $$\liminf_{x\to\infty} \frac{\log\frac{1}{x}\log\frac{1}{T(x)}}{\log\log x} \geq 1/\alpha.$$

To show that the inequality sign cannot hold we need the following lemma.

Lemma 7.2.1. *If a distribution function $F(x)$ has the property that*
$$T(x) \leq L \exp[-\lambda x(\log x)^\delta],$$
where L, λ and δ are positive constants, then the characteristic function $f(t)$ of $F(x)$ is an entire function such that

(7.2.2) $\quad \lambda = \limsup_{r\to\infty} [\log\log\log M(r,f)/\log r] \leq 1/\delta.$

The number λ is sometimes called the form of $f(t)$.

The proof of this lemma may be found in Ramachandran (1962).

We use an indirect proof to show that the inequality sign cannot hold in (7.2.1). Suppose therefore that the inequality sign holds. Then there exists a number $\gamma < \alpha$ such that
$$\lambda = \limsup_{r\to\infty} [\log\log\log M(r,f)/\log r] \geq 1/\gamma$$
for sufficiently large x. This means that
$$T(x) \leq \exp[-x(\log x)^{(1/\gamma)}].$$
It follows from Lemma 7.2.1 that
$$\limsup_{r\to\infty} [\log\log\log M(r,f)/\log r] \leq \gamma.$$
Since $\gamma < \alpha$ this is a contradiction, so that the inequality sign cannot hold in (7.2.1) and (ii) is necessary.

We next show the sufficiency of the conditions. In view of (i), $F(x)$ does not have an entire characteristic function of order 1 and exponential type, and it follows that
$$\frac{\log\left[\frac{1}{x}\log\frac{1}{T(x)}\right]}{\log\log x}$$
is defined for all sufficiently large x. Condition (ii) implies that

$$T(x) \leqslant \exp\left[-x(\log x)^{(1/\alpha)+\epsilon}\right]$$

for all x sufficiently large, i.e. $x \geqslant X(\epsilon)$. It follows from Lemma 7.2.1 that

$$\limsup_{r \to \infty} \left[\log \log \log M(r, f)/\log r\right] \leqslant \alpha.$$

It can then again be shown that the inequality sign cannot hold in the last relation, so that the theorem is proved.

The order and type of entire functions provide means of studying their growth. This study can be refined by introducing proximate orders and types with respect to proximate orders [see e.g. B. Ya. Levin (1964), pp. 31 ff.].

The definitions of proximate orders with statements concerning some of their properties are given in Appendix A1.

Next we investigate the behaviour of distribution functions which have entire characteristic functions with a given proximate order.

We consider first entire characteristic functions which have one-sided distribution functions $H(x)$ with $H(0) = 0$.

Theorem 7.2.2. Let $H(x)$ be a distribution function which admits the representation

(7.2.3) $1 - H(x) = \exp\{-kx^{1+\rho(x)}\}$ $(x > 0)$,

where $\rho(x)$ is a proximate order while k is a constant. Then $H(x)$ has an entire characteristic function $h(z)$ such that

$$\lim_{r \to \infty} \frac{\log M(r, h)}{r\bar{\rho}(r) + 1} = k^{-1/\rho} c,$$

with $c = \dfrac{\rho}{(1 + \rho)^{1 + 1/\rho}}$ (where ρ is the order of $h(z)$) and $\bar{\rho}(z)$ is a dual proximate order to $\rho(x)$).

Proof. Since $h(z) = \displaystyle\int_0^\infty e^{izu}\, dH(u)$, we see (integrating by parts) that

$$h(z) = 1 + iz \int_0^\infty e^{izu}[1 - H(u)]\, du.$$

It follows that

(7.2.4) $M(r, h) = \max\left[h(ir), h(-ir)\right] = 1 + r \displaystyle\int_0^\infty e^{ru}[1 - H(u)]\, du.$

Let [see property (5) of proximate orders in Appendix A1]

(7.2.5) $f(r, u) = e^{ru}[1 - H(u)]$.

ANALYTIC CHARACTERISTIC FUNCTIONS

Then we see that

$$M(r, h) = 1 + r \int_0^\infty e^{-u} f(r+1, u) \, du$$

and

(7.2.6) $\quad M(r, h) \leqslant 1 + rf(r+1, u(r+1)).$

For the sake of simplicity we write u_1 instead of $u(r+1)$ and have

$$M(r, h) \leqslant 1 + rf(r+1, u_1).$$

After an elementary computation we obtain

(7.2.7) $\quad u_1^{\rho(u_1)} = \dfrac{1}{k} \dfrac{r+1}{1 + \rho(u_1) + \rho'(u_1) u_1 \log u_1}$

and obtain, in view of (7.2.1) and (7.2.4),

(7.2.8) $\quad f(r+1, u_1) = \exp\left\{(r+1) u_1 \dfrac{[\rho(u_1) + \rho'(u_1)] u_1 \log u_1}{1 + \rho(u_1) + \rho'(u_1) u_1 \log u_1}\right\}.$

Let $x = \phi(r)$ be the inverse function to $x^{\rho(x)}$, that is

$$[\phi(r)]^{\rho[\phi(r)]} = r.$$

We see then from (7.2.6) that

$$\dfrac{u_1^{\rho(u_1)}}{[\phi(r)]^{\rho[\phi(r)]}} = \dfrac{1}{k} \dfrac{r+1}{r} \dfrac{1}{1 + \rho(u_1) + \rho'(u_1) u_1 \log u_1}.$$

Hence (note that u_1 is a function of r)

$$\lim_{r \to \infty} \dfrac{u_1^{\rho(u_1)}}{[\phi(r)]^{\rho[\phi(r)]}} = \dfrac{1}{k} \dfrac{1}{1+\rho}$$

so that

(7.2.9) $\quad \lim_{r \to \infty} \dfrac{u_1(r)}{\phi(r)} = k^{-1/\rho} \dfrac{1}{(1+\rho)^{1/\rho}}.$

It follows from (7.2.8) and (7.2.7) that

$$\dfrac{\log M(r, h)}{r^{\bar\rho(r)+1}} \leqslant \dfrac{r+1}{r} \dfrac{u_1}{r^{\bar\rho(r)}} \dfrac{\rho(u_1) + \rho'(u_1) u_1 \log u_1}{1 + \rho(u_1) + \rho'(u_1) u_1 \log u_1} + o(1)$$

as $r \to \infty$. It follows from (7.2.9) that

$$\limsup_{r \to \infty} \dfrac{\log M(r, h)}{r^{\bar\rho(r)+1}} \leqslant \dfrac{\rho}{(1+\rho)^{1+1/\rho}} k^{-1/\rho}.$$

It was shown in Rossberg (1967) that

$$\liminf_{r \to \infty} \frac{\log M(r, h)}{r^{\bar{\rho}(r)+1}} \geq \frac{\rho}{(1+\rho)^{1+1/\rho}} k^{-1/\rho}.$$

The last two inequalities imply the statement of the theorem.

Corollary to Theorem 7.2.2. Let $\bar{\rho}(r)$ be a proximate order. Then there exists a distribution function $H(x)$ whose characteristic function $h(z)$ is an entire function which has a proximate order $1 + \bar{\rho}(r)$ such that (7.2.2) holds.

To prove the corollary we let $\rho(r)$ be the dual proximate order to $\bar{\rho}(r)$ and construct the distribution function $H(x)$ according to (7.2.1).

Remark. If $1 - H(x) \leq \exp[-kx^{1+\rho(x)}]$ for sufficiently large x then the characteristic function $h(z)$ of $H(x)$ is an entire function of order not exceeding $1 + 1/\rho$ and

$$\overline{\lim_{r \to \infty}} \; [\log M(r, h)]/r^{\bar{\rho}(r)+1} \leq ck^{-1/\rho},$$

where $c = \rho/[1+\rho]^{1+1/\rho}$. This follows from the proof of Theorem 7.2.1.

The growth of an entire characteristic function which belongs to a non-finite distribution function $H(x)$ with lext $H = 0$ depends on the tail of the function $H(x)$ which can be measured by

$$T(H, x) = \log\{-\log[1 - H(x)]\}/\log x - 1.$$

Theorem 7.2.3. A distribution function $H(x)$ for which $H(0) = 0$ has an entire characteristic function $h(z)$ such that

$$(7.2.10) \quad 0 < \limsup_{r \to \infty} \frac{\log M(r, h)}{r^{\bar{\rho}(r)+1}} = \bar{\sigma} < \infty$$

[*i.e., $h(z)$ has proximate order $\bar{\rho}(r) + 1$*] if, and only if,

(i) $1 - H(x) > 0$ for every $x > 0$,

(ii) $\liminf_{r \to \infty} x^{T(H,x) - \rho(x)} = (c/\bar{\sigma})^{\rho} = k$ (say).

We first prove that these conditions are sufficient.
We see from (ii) that for $\epsilon > 0$

$$x^{T(H,x) - \rho(x)} > k - \epsilon,$$

so that $x^{T(H,x)+1} > (k-\epsilon) x^{\rho(x)+1}$. We use the definition of $T(H, x)$ and obtain easily

$$[1 - H(x)] < \exp[-(k-\epsilon) x^{\rho(x)+1}].$$

ANALYTIC CHARACTERISTIC FUNCTIONS

Hence $h(z)$ is an entire function and

$$\limsup_{r \to \infty} [\log M(r, h)]/r^{\bar{\rho}(r)+1} \leq \bar{\sigma}.$$

We use a result of Rossberg [(1967), theorem 4] and see that

$$\limsup_{r \to \infty} \frac{\log M(r, h)}{r^{\bar{\rho}(r)+1}} \geq \bar{\sigma}.$$

(7.2.10) follows from the last two inequalities.

The necessity of the conditions follows easily from a result of Rossberg [(1967), theorem 8].

M. Dewess and M. Riedel (1977) obtained similar results for the case where

$$\lim_{r \to \infty} \frac{\log M(r, h)}{r^{\bar{\rho}(r)+1}} = 0$$

and also for the case where

$$\limsup_{r \to \infty} \frac{\log M(r, h)}{r^{\bar{\rho}(r)+1}} = \infty.$$

In order to generalize some of the preceding results we introduce the following definitions.

We say that the function $\beta(x)$, $0 \leq x \leq \infty$ belongs to the class N if it is positive, differentiable, and if it increases monotonically to $+\infty$ and if it satisfies the condition

$$(7.2.11) \quad \lim_{x \to \infty} \frac{\beta[(1 + \lambda(x))x]}{\beta(x)} = 1$$

for any function $\lambda(x)$ which tends monotonically to zero as $x \to \infty$.

The function $\gamma(x)$ is said to belong to the class $\wedge$ if it belongs to the class N and if it satisfies the condition

$$(7.2.12) \quad \lim_{x \to +\infty} \frac{\gamma(cx)}{\gamma(x)} = 1$$

for any constant c such that $0 < c < \infty$. This limit is uniform with respect to c for

$$0 < c_1 \leq c \leq c_2 < +\infty.$$

Theorem 7.2.4. *Let $\gamma(x)$ and $\beta(x)$ be functions which belong to the class $\wedge$ and N respectively. The function $F(x)$ has a characteristic function such that*

$$\limsup_{r \to \infty} \frac{\gamma[(1/r) \log M(r, h)]}{\beta(\log r)} = 1/\alpha \quad (\alpha > 0)$$

if, and only if,

$$\liminf_{r \to +\infty} \frac{\beta\left[\log\left(\frac{1}{x}\log\frac{1}{T(x)}\right)\right]}{\gamma(x)} = \alpha.$$

Theorems 7.2.4 and 7.2.1 are particular cases of the last theorem. By putting $\gamma(x) = \log x$ and $\beta(x) = x$ we obtain Theorem 7.2.4; similarly we obtain Theorem 7.2.1 by substituting $\gamma(x) = \log \log x$ and $\beta(x) = x$ in Theorem 7.2.4.

For the proof of Theorem 7.2.4 we refer the reader to N. I. Yakovleva (1972). Yakovleva (1976) also studied the connection between the tail behaviour of a distribution function and the lower order of its characteristic function (see Appendix A1). This author also showed that for arbitrary ρ and λ such that $1 \leqslant \lambda \leqslant \rho \leqslant \infty$ there exists an entire characteristic function whose order equals ρ while its lower order is λ.

We also mention, without proof, a result of B. V. Vinnickii (1975) on entire characteristic functions of infinite order.

Let $\Phi(x) = \log M(e^x)$ and[*] denote by $\Phi^0(x)$ the inverse of $\Phi(x)$.

Theorem 7.2.5. Suppose that the entire characteristic function $f(t)$ has infinite order. Then

$$\liminf_{x \to \infty} \log\left[\frac{1}{x}\log\frac{1}{T(x)}\right] \Big/ \Phi^0(x) = 1.$$

7.3 Entire characteristic functions of order 2 with a finite number of zeros

Finally we consider the family of entire characteristic functions of order 2 which have only a finite number of zeros.[†] Denote this family by $\mathfrak{G}_2$. This study is motivated by the fact that $\mathfrak{G}_2$ is one of the few families whose decomposition properties are known. According to Hadamard's factorization theorem the functions of $\mathfrak{G}_2$ have the form

$$f(t) = P(t)\, e^{A(t)},$$

where $P(t)$ and $A(t)$ are polynomials and $A(t)$ is of second degree. In view of the Hermitian property of characteristic functions the polynomial $P(t)$ is of even degree and has the form

(7.3.1) $$P(t) = \prod_{j=1}^{n}\left(1 - \frac{t}{\zeta_j}\right)\left(1 + \frac{t}{\bar{\zeta}_j}\right)$$

---

[*] $M(r)$ is, as usual, the maximum modulus of $f(x)$ in the circle $|z| \leqslant r$; $T(x)$ is the tail of the corresponding distribution function.

[†] These are necessarily located symmetrically with respect to the imaginary axis.

ANALYTIC CHARACTERISTIC FUNCTIONS

where ζ_j and $\bar{\zeta}_j$ are the zeros of $P(t)$. Alternatively one can expand $P(t)$ and write

$$P(t) = \sum_{j=0}^{2m} \lambda_j (it)^j,$$

where the λ_j are real and $\lambda_0 = 1$. Since we are only interested in the decompositions of functions belonging to $\mathfrak{G}_2$ we can discard functions $e^{i\beta t}$ (β real) and write

(7.3.1a) $\quad f(t) = e^{-\sigma^2 t^2/2} P(t).$

We introduce the Hermite polynomials

(7.3.2) $\quad H_k(x) = e^{x^2/2} \dfrac{d^k}{dx^k} (e^{-x^2/2})$

and see that

$$\frac{d^k}{dt^k}(e^{-t^2/2}) = H_k(t) e^{-t^2/2}.$$

It is not difficult to compute the coefficients of the polynomials $H_k(x)$ [see G. Szegö (1959), pp. 104-5]. One obtains

(7.3.3) $\quad H_k(x) = (-1)^k 2^{-k/2} k! \sum_{j=0}^{[k/2]} \dfrac{(-1)^j}{j!} \dfrac{(x\sqrt{2})^{k-2j}}{(k-2j)!}$

We then see that

(7.3.4) $\quad \begin{cases} h_{2k}(t) = T_2^k e^{-t^2/2} = \dfrac{(-1)^k}{\alpha_{2k}} H_{2k}(t) e^{-t^2/2} \\[2mm] h_{2k-1}(t) = T_1 T_2^{k-1} e^{-t^2/2} = \dfrac{(-1)^k}{\alpha_{2k}} \dfrac{H_{2k-1}(t)}{t} e^{-t^2/2} \end{cases}$

for $k = 1, 2, \ldots$.

Here T_1 and T_2 are operators on characteristic functions $f(t)$ which belong to distribution functions having finite second moments. They are defined in the following way

$$T_1 f(t) = \frac{f'(t) - f'(0)}{t f''(0)}$$

$$T_2 f(t) = \frac{f''(t)}{f''(0)}.$$

It is known that $T_1 f(t)$ and $T_2 f(t)$ are again characteristic functions.

We write T_2^k for the k-times iterated operator T_2. The number

$$\alpha_{2k} = (-1)^k H_{2k}(0) = (2k)!/(2^k k!)$$

is the moment of order $2k$ of the normal distribution with zero mean and variance one.

The function (7.3.2) is a characteristic function, so that we can use the inversion formula to derive the corresponding frequency function. One easily sees that

$$p(x) = \frac{1}{\sigma\sqrt{2\pi}} Q(x) e^{-x^2/2\sigma^2},$$

where

(7.3.5) $\quad Q(x) = \sum_{k=0}^{2m} (-1)^k \lambda_k \sigma^{-k} H_k(x/\sigma).$

Since $f(t)$ is a characteristic function the polynomial $Q(x)$ is necessarily non-negative for all real x. $Q(x)$ is called the associated polynomial of $f(t)$.

It follows from formula (7.3.1a) that all factors of a characteristic function which belong to $\mathfrak{G}_2$ are also in $\mathfrak{G}_2$. Moreover, the distribution functions of these factors are absolutely continuous.

A function $f(t) \in \mathfrak{G}_2$ has the form (7.3.1a) and there are two possibilities: either $f(t)$ is indecomposable or it admits a decomposition

(7.3.6) $\quad f(t) = f_1(t) f_2(t).$

There are again two possibilities in the case where $f(t)$ is decomposable. In this case $f(t)$ may have a normal factor $f_1(t) = \exp(-\sigma_1^2 t^2/2)$; then the second factor is

$$f_2(t) = P(t) \exp[-\sigma_2^2 t^2/2],$$

where $\sigma_1^2 + \sigma_2^2 = \sigma^2$; or both factors $f_1(t)$ and $f_2(t)$ have the form (7.3.1a); that is

$$f_j(t) = P_j(t) \exp[-\sigma_j^2 t^2/2] \in \mathfrak{G}_2 \qquad (j = 1, 2).$$

Here $\sigma_1^2 + \sigma_2^2 = \sigma^2$ and $P_1(t) P_2(t) = P(t)$.

We consider first the case where $f(t)$ is decomposable so that at least one of its factors, say $f_2(t)$, has the form (7.3.1a).

Theorem 7.3.1. *Suppose that the characteristic function* $f(t) \in \mathfrak{G}_2$ *admits a non-trivial decomposition; then its associated polynomial $Q(x)$ has no real zeros.*

Let $p(x)$ and $p_2(x)$ be the frequency functions of $f(t)$ and $f_2(t)$ respectively. Then

(7.3.7) $\quad p(x) = \dfrac{1}{\sigma\sqrt{2\pi}} Q(x) \exp[-x^2/(2\sigma^2)]$

(7.3.7a) $$p_2(x) = \frac{1}{\sigma_2\sqrt{2\pi}} Q_2(x) \exp[-x^2/(2\sigma_2^2)],$$

where $Q(x)$ and $Q_2(x)$ are non-negative polynomials determined by the zeros of $f(t)$ and $f_2(t)$ respectively. It follows from (7.3.7), (7.3.7a) and (7.3.6) that

$$p(x) = \int_{-\infty}^{\infty} p_2(x-y) \, dF_1(y)$$

$$= \frac{1}{\sigma_2\sqrt{2\pi}} \int_{-\infty}^{\infty} Q_2(x-y) \exp[-(x-y)^2/(2\sigma_2^2)] \, dF_1(y).$$

We give an indirect proof for the theorem and assume tentatively that the associated polynomial $Q(x)$ has a real zero x_0. Then $p(x_0) = 0$ so that

(7.3.8) $$\int_{-\infty}^{\infty} Q_2(x_0-y) \exp[-(x_0-y)^2/(2\sigma_2^2)] \, dF_1(y) = 0.$$

Since $Q_2(x)$ is non-negative we see that the integrand in (7.3.8) cannot vanish. Hence (7.3.8) can be satisfied only if $F_1(y)$ is a purely discrete distribution whose discontinuity points are the zeros of $Q_2(x_0-y)$. However, all factors of $f(t)$ belong to $\mathfrak{G}_2$ and are therefore continuous. This contradiction proves Theorem 7.3.1.

Corollary. *The characteristic functions*

$$h_{2k}(t) = \frac{(-1)^k}{\alpha_{2k}} H_{2k}(t) \, e^{-t^2/2}$$

are indecomposable.

We differentiate the relation

$$e^{-t^2/2} = \frac{1}{\sqrt{2\pi}} \int_{-\infty}^{\infty} \exp(ity - y^2/2) \, dy$$

($2k$) times and obtain

$$H_{2k}(t) \, e^{-t^2/2} = \frac{(-1)^k}{\sqrt{2\pi}} \int_{-\infty}^{\infty} y^{2k} \exp(ity - y^2/2) \, dy.$$

The frequency function which belongs to $h_{2k}(t)$ is then

$$p_{2k}(x) = (1/\alpha_{2k}) x^{2k} \exp(-x^2/2),$$

so that $p_{2k}(0) = 0$. We conclude from Theorem 7.3.1 that $h_{2k}(t)$ is indecomposable.

In order to derive a converse to Theorem 7.3.1 we need an explicit expression for the polynomial $Q(x)$. Combining (7.3.3) and (7.3.5), we get, after some elementary computations,

$$(7.3.9) \quad \sigma^{2m} Q(x) = \sum_{k=0}^{2m} c_{m,k}(\sigma)(x/\sigma)^k = S(x/\sigma)$$

where

$$c_{m,k}(z) = \frac{1}{k!} \sum_{j=0}^{m-[(k+1)/2]} \frac{(-1)^j}{2^j j!} \lambda_{k+2j}(k+2j)! \, z^{2m-k-2j}.$$

Let $f(t) \in \mathfrak{G}_2$; then, in view of (7.3.1a),

$$f(t) = e^{-\sigma^2 t^2/2} \sum_{j=0}^{2m} \lambda_j (it)^j.$$

According to (7.3.7) and (7.3.9) the corresponding frequency function is

$$(7.3.10) \quad p(x) = \sigma^{-2m-1}(2\pi)^{-1/2} e^{-x^2/(2\sigma^2)} S(x/\sigma).$$

Next we prove the following statement which is slightly stronger than the converse of Theorem 7.3.1.

Theorem 7.3.2. *Let $f(t) \in \mathfrak{G}_2$ and suppose that the polynomial associated with $f(t)$ has a real zero; then $f(t)$ has a normal factor.*

The function $f(t)$, given by (7.3.1a), has a normal factor if, and only if, it can be written in the form

$$f(t) = n_\theta(t) f_\theta(t)$$

where

$$n_\theta(t) = \exp[-\sigma^2(1-\theta^2) t^2/2] \quad (0 < \theta < 1)$$

and where

$$f_\theta(t) = \exp(-\sigma^2 \theta^2 t^2/2) \sum_{j=0}^{2m} \lambda_j (it)^j$$

is a characteristic function. The existence of a normal factor can be established by showing that $f_\theta(t)$ is a characteristic function for some $\theta \in (0, 1)$. If this is the case then the function

$$p_\theta(x) = \frac{1}{2\pi} \int_{-\infty}^{\infty} e^{-itx} f_\theta(t) \, dt$$

is real and non-negative for all real x. Replacing in (7.3.10) and (7.3.9) σ by $\theta\sigma$, we obtain

$$p_\theta(x) = (\theta\sigma)^{-2m-1}(2\pi)^{-1/2} e^{-x^2/(2\theta^2\sigma^2)} \sum_{k=0}^{2m} c_{m,k}(\theta\sigma)(x/\theta\sigma)^k.$$

This is a frequency function if and only if

$$S_\theta(y) = \sum_{k=0}^{2m} c_{m,k}(\theta\sigma) y^k$$

is non-negative for all real y.

We note that $p_1(x) = p(x)$ and $S_1(y) = S(y)$. It follows from the assumption of the theorem that $S_1(y) = S(y)$ for all real y. Since

$$S_1(y) = \sum_{k=0}^{2m} c_{m,k} y^k$$

is a polynomial of degree $2m$ we conclude that $c_{m,2m} > 0$.

Let η_j ($j = 1, 2, \ldots, r$; $r \leq 2m$) be the zeros of $S_1(y)$ and suppose that η_j has the multiplicity n_j. Then

$$S(y) = c_{m,2m} \prod_{j=1}^{r} (y - \eta_j)^{n_j}$$

We select $\rho > 0$ so that $2\rho < \min |\mathrm{Im}\, \eta_j|$ and denoting the circle $|z - \eta_j| < \rho$ by C_ρ, we write

$$G = \bigcup_{j=1}^{r} C_j.$$

Using the fact that the roots of a polynomial are continuous functions of the coefficients, it can be shown that there exists a θ_1 such that the polynomial

$$\sum_{k=0}^{2m} c_{m,k}(\theta_1 \sigma) \left(\frac{x}{\theta_1 \sigma}\right)^k > 0$$

for all real x. This means that $p_{\theta_1}(x)$ is a frequency function and that $f(t)$ has the normal factor $n_{\theta_1}(t)$. This completes the proof of Theorem 7.3.2.

Combining Theorems 7.3.1 and 7.3.2, one obtains the following statement:

Theorem 7.3.3. A characteristic function $f(t) \in \mathfrak{G}_2$ is indecomposable if and only if its associated polynomial has at least one real zero.

Finally we mention briefly a few additional results concerning the class $\mathfrak{G}_2$. For details and proofs we refer to Lukacs (1967).

(i) Let $f(t)$ be a decomposable characteristic function of $\mathfrak{G}_2$ whose associated polynomial has no real zero. Then it is possible to determine its normal component with maximal variance.

(ii) Let $f_1(t) \in \mathfrak{G}_2$ and $f_2(t) \in \mathfrak{G}_2$ and suppose that both are indecomposable. Then $f(t) = f_1(t) f_2(t) \in \mathfrak{G}_2$ and the associated polynomial of $f(t)$ has no real zeros. In view of Theorem 7.3.2, $f(t)$ has a normal factor. This shows that the product of two indecomposable characteristic functions can have a normal factor even if none of its components admits a normal factor. A different example, not involving functions of $\mathfrak{G}_2$, was given by R. A. Fisher & D. Dugué (1968).

(iii) The characteristic functions defined by (7.3.4) are decomposable and have normal factors.

(iv) It is also possible to ascertain whether a characteristic function $f(t) = P(t) \exp(-\sigma^2 t^2/2)$ of $\mathfrak{G}_2$ admits a decomposition of the form

$$f(t) = f_1(t) f_2(t) \quad \text{where}$$

$$f_j(t) = P_j(t) \exp(-\sigma_j^2 t^2/2) \in \mathfrak{G}_2 \quad \text{and where}$$

$$P(t) = P_1(t) P_2(t); P_j(t) \not\equiv 1, \quad P_1(0) = P_2(0) = 1,$$

while $\sigma^2 = \sigma_1^2 + \sigma_2^2$.

7.4 Special families of analytic and entire characteristic functions

In this section we discuss analytic characteristic functions which are defined by some special properties.

In a conversation, D. van Dantzig mentioned a class of characteristic functions which had been introduced in the prize questions of the *Nieuw Archief voor Wiskunde* (1958-60).

Let $f(z)$ be an analytic characteristic function which has the strip of regularity $-\alpha < \operatorname{Im} z < \beta$. Here $\alpha > 0$, $\beta > 0$ may also assume the value $+\infty$. Then $f(it)$ is defined, provided $-\alpha < t < \beta$.

An analytic characteristic function f is said to belong to the class $\mathfrak{D}$ (van Dantzig class) if, for $\alpha < t < \beta$,

$$g(t) = \frac{1}{f(it)}$$

is also a characteristic function. Pairs $[f, g]$ of characteristic functions of $\mathfrak{D}$ are connected by the relation

(7.4.1) $\quad g(t) f(it) = 1$.

It is obvious that the pairs

$$\left[\cos t, \frac{1}{\cosh t}\right], \quad \left[\frac{\sin t}{t}, \frac{t}{\sinh t}\right], \quad [e^{-t^2/2}, e^{-t^2/2}]$$

belong to $\mathfrak{D}$. The last example shows that there exist characteristic functions such that

(7.4.2) $\quad f(t) = \dfrac{1}{f(it)}$.

ANALYTIC CHARACTERISTIC FUNCTIONS 93

We call such pairs self-reciprocal and denote the set of self-reciprocal pairs by $\mathbb{D}_s$. Clearly $\mathbb{D}_s \subset \mathbb{D}$.

Let $F(x)$ and $G(x)$ be two distribution functions such that the pair $[f(t), g(t)] \in \mathbb{D}$. Then

$$g(t) = \frac{1}{f(it)}$$

while

$$f(it) = \int_{-\infty}^{\infty} e^{-tx} \, dF(x), \qquad g(-it) = \int_{-\infty}^{\infty} e^{tx} \, dG(x).$$

Therefore $f(t)$ and $g(t)$ are both real and $[g(t), f(t)] \in \mathbb{D}$.

It is also easily seen that the class $\mathbb{D}$ is closed under multiplications and that $f(t) \in \mathbb{D}$ implies

$$\frac{f(t)}{f(it)} \in \mathbb{D}_s.$$

We write $\phi(t) = \log f(t)$, $\gamma(t) = \log g(t)$ and denote the cumulants of $F(x)$ and $G(x)$ respectively by κ_i and λ_i respectively. Here $F(x)$ and $G(x)$ are the distribution functions corresponding to $f(t)$ and $g(t)$. Then

$$\gamma(t) + \phi(it) = 0,$$

while

(7.4.3) $\quad \kappa_{2v-1} = \lambda_{2v-1} = 0$

$\qquad \kappa_{2v} + (-1)^v \lambda_{2v} = 0.$

Hence we obtain the following result:

Theorem 7.4.1. If a characteristic function $f(t) \in \mathbb{D}$ then it is real and even, and if $g(t) = 1/f(it)$ then $g(t) \in \mathbb{D}$ and the cumulants of $f(t)$ and $g(t)$ satisfy the relations (7.4.3). The family $\mathbb{D}$ is closed under multiplications and $f(t) \in \mathbb{D}$ implies $f(t)/f(it) \in \mathbb{D}_s$.

The statement about the cumulants follows easily from the power-series expansions of $\phi(t)$ and $\gamma(t)$.

Corollary to Theorem 7.4.1. If an infinitely divisible characteristic function $f(t)$ belongs to $\mathbb{D}$ then all its cumulants are non-negative.

According to the assumptions of the corollary, the characteristic function $f(x)$ admits the Kolmogorov canonical representation

$$\log f(t) = \phi(t) = \int_{-\infty}^{\infty} \frac{\cos tx - 1}{x^2} \, dK(x),$$

where $K(x)$ is a non-decreasing function such that $K(-\infty) = 0$ and $K(+\infty) = \kappa_2$. Differentiating the canonical representation, one sees that

$$g(t) = \frac{\phi''(t)}{\phi''(0)} = \frac{1}{\kappa_2} \int_{-\infty}^{\infty} \cos tx \, dK(x)$$

is a characteristic function. It follows from the power-series expansion of $\phi(t)$ that

$$g(t) = \frac{1}{\kappa_2} \sum_{v=1}^{\infty} (-1)^{v-1} \frac{\kappa_{2v}}{(2v-2)!} t^{2v-2}.$$

Let α_v be the moment of order v of the distribution function $\kappa_2^{-1} K(x)$; then $\alpha_{2v-2} = \kappa_{2v}/\kappa_2$ ($v = 1, 2, \ldots$) while $\alpha_{2v-1} = 0$. The statement of the corollary follows from the fact that moments of even order are non-negative.

Theorem 7.4.2. *Let $f_1(t), f_2(t), \ldots, f_n(t)$ be n arbitrary different characteristic functions from $\mathfrak{D}_s$, and let $p_j > 0$ ($j = 1, 2, \ldots, n$) be n positive numbers such that*

$$\sum_{j=1}^{n} p_j = 1.$$

Then the characteristic function

(7.4.4) $\quad h(t) = \sum_{j=1}^{n} p_j f_j(t)$

is not self-reciprocal.

We give an indirect proof and assume that the $f_j(t) \in \mathfrak{D}_s$ and that $h(t)$ also belongs to $\mathfrak{D}_s$. We see from (7.4.1) that

$$f_j(t) f_j(it) = 1, \qquad h(t) h(it) = 1.$$

Therefore

$$\left[\sum_{j=1}^{n} p_j f_j(t) \right] \left[\sum_{k=1}^{n} \frac{p_k}{f_k(t)} \right] = 1.$$

Since $\left[\sum_{j=1}^{n} p_j \right]^2 = 1$ we have

$$\sum_{\substack{j=1 \\ j \neq k}}^{n} \sum_{k=1}^{n} p_j p_k \frac{f_j}{f_k} = \sum_{j=1}^{n} \sum_{k=1}^{n} p_j p_k$$

ANALYTIC CHARACTERISTIC FUNCTIONS

or
$$\sum_{\substack{j,k=1 \\ j \neq k}}^{n} p_j p_k \frac{(f_j - f_k)}{f_k} = 0.$$

We write here, and in the preceding formula, f_j and f_k instead of $f_j(t)$ and $f_k(t)$. This sum contains with each term

$$p_j p_k (f_j - f_k)/f_j$$

also the term $p_k p_j (f_k - f_j)/f_k$. We join these terms and see that

(7.4.5) $$\sum_{j=1}^{n-1} \sum_{k=j+1}^{n} \frac{p_j p_k}{f_j f_k} (f_j - f_k)^2 = 0.$$

If t is sufficiently small then the functions $f_j(t)$ are all positive and we can conclude from (7.4.5) that all the functions $f_j(t)$ are equal. This contradiction completes the proof of Theorem 7.4.2.

Theorem 7.4.3. The only infinitely divisible characteristic function which is self-reciprocal is the characteristic function of the normal distribution with zero mean.

Let $f(t)$ be infinitely divisible and suppose that $f \in \mathbb{B}_s$. In view of Theorem 7.4.1, $f(t)$ is even and admits the canonical representation

$$\phi(t) = \log f(t) = \int_{-\infty}^{\infty} (\cos tx - 1) \frac{dK(x)}{x^2}.$$

By assumption, $\phi(t) + \phi(it) = 0$, so that

(7.4.6) $$\int_{-\infty}^{\infty} (\cos tx + \cosh tx - 2) \frac{dK(x)}{x^2} = 0.$$

The function $\cos z + \cosh z - 2$ is non-negative for all real z, so that (7.4.6) implies $K(x) = \sigma^2 \in (x)$. It follows immediately that $f(t) = \exp[-\sigma^2 t^2/2]$.

We again use two operators whose domains are characteristic functions $f(t)$ which have finite second moments. Let

$$T_1 f(t) = \frac{f'(t) - f'(0)}{t f''(0)}$$

$$T_2 f(t) = \frac{f''(t)}{f''(0)}.$$

If $f(t)$ is an analytic characteristic function then $T_1 f(t)$ and $T_2 f(t)$ are also analytic characteristic functions.

We consider even, entire characteristic functions which have only real zeros $\{t_j\}$ and have the order $\rho \leq 2$ and the exponent of convergence $\rho_1 < 2$. We denote this class of characteristic functions by $\mathfrak{D}_E$.

It follows from Hadamard's factorization theorem [Appendix A1, Theorem A3] and from $f(0) = 1$ that

$$f(z) = e^{-cz^2} \prod_j \left(1 - \frac{z^2}{t_j^2}\right)$$

where $c > 0$ if $\rho = 2$ but $c = 0$ if $\rho < 2$. Then

$$\frac{1}{f(it)} = e^{-ct^2} \prod_j \left(1 + \frac{t^2}{t_j^2}\right)^{-1} \quad (t \text{ real})$$

is also a characteristic function, so that $f(t) \in \mathfrak{D}$. Thus $\mathfrak{D}_E \subset \mathfrak{D}$.

Let $f(z) \in \mathfrak{D}_E$. The function $f(z)$ satisfies the conditions of Theorem A4 in Appendix A1; therefore $f'(z)$ is an entire function which has only real zeros, and $f'(z)$ has order $\rho \leq 2$ and exponent of convergence $\rho_1 < 2$. Moreover $z = 0$ is a simple zero of $f'(z)$ and $f'(z)/z$ is even, so that $T_1 f(t) \in \mathfrak{D}_E$. The function $f'(z)$ also satisfies the conditions of Theorem A4, so that $f''(z)$ is an entire function of order ρ and an exponent of convergence ρ_1. The function $f''(z)$ has only real zeros and $f''(0) \neq 0$. Therefore $T_2 f(t) \in \mathfrak{D}_E$ and we have obtained the following result:

Theorem 7.4.4. *The set* $\mathfrak{D}_E \subset \mathfrak{D}$ *is closed under the application of the operators* T_1 *and* T_2 *and their iterates.*

The operators T_1 and T_2 can be used to generate functions of $\mathfrak{D}$. A number of examples can be found in Lukacs (1968) where also properties of frequency functions which belong to characteristic functions from $\mathfrak{D}$ are discussed.

I. V. Ostrovskii (1970a) has also studied the class $\mathfrak{D}$. We mention here some of his results.

Theorem 7.4.5. *Every characteristic function* $f(t) \in \mathfrak{D}$ *is meromorphic in some cross*

$$K(a, b) = [|\operatorname{Re} t| < a] \cup [|\operatorname{Im} t| < b],$$

where $a > 0, b > 0$.

Since $f(t)$ is an analytic function in some neighbourhood of the origin and $f(t) \in \mathfrak{D}$, we see that $1/f(it)$ is also analytic in a neighbourhood and conclude from Theorem 7.4.1 the statement of the theorem.

Ostrovskii also showed that there exists a characteristic function $f(t) \in \mathfrak{D}$ which is meromorphic in the cross $K(\pi, \pi)$ for which every point on the boundary of $K(\pi, \pi)$ is a singular point.

ANALYTIC CHARACTERISTIC FUNCTIONS

He also studied the regions in which functions of $\mathfrak{D}$ are meromorphic and obtained estimates for functions of $\mathfrak{D}$. We give two examples of his estimates:

Theorem 7.4.6. Every $f(t) \in \mathfrak{D}$ admits for some $c > 0$ the estimate

$$M(r,f) \leqslant \exp(e^{cr}), \quad r \geqslant 1 \quad [M(r,f) = \max_{|t|=r} |f(t)|].$$

There exist entire characteristic functions $f(t) \in \mathfrak{D}$ such that

$$M(r,f) \geqslant \exp(e^{cr}), \quad r \geqslant 1$$

for some $c > 0$.

Theorem 7.4.7. Let $f(t)$ be an entire function and denote the zeros of $f(t)$ by $\{a_k\}$. Suppose that $f(t) \in \mathfrak{D}$ and $0 < R < \min_k |a_k|$. Then

$$M(r,f) \leqslant \exp\left\{C \exp\left(\frac{\pi}{R}r\right)\right\} \quad (r \geqslant 1),$$

where C is a finite positive constant which does not depend on r. If $f(t)$ has no zeros then R is an arbitrary positive number.

D. Dugué [(1957), p. 21] noticed that there exist characteristic functions such that

(7.4.7) $\quad \frac{1}{2}[f_1(t) + f_2(t)] = f_1(t) f_2(t).$

As an example he mentioned the pair of characteristic functions

(7.4.7a) $\quad f_1(t) = (1 + it)^{-1}, \quad f_2(t) = (1 - it)^{-1}$

and he posed the problem of finding other characteristic functions satisfying (7.4.7). The functions (7.4.7a) are not the only pair which satisfies (7.4.7). Let

(7.4.7b) $\quad \begin{cases} f_1(t) = \frac{1}{2}[1 + \cos tb - i \sin tb] \\ f_2(t) = \frac{1}{2}[1 + \cos tb + i \sin tb]; \end{cases}$

then it is easily seen that these are characteristic functions which satisfy (7.4.7).

Remark. If $f(t)$ is a characteristic function such that

$$|f(t)|^2 = \operatorname{Re} f(t)$$

then

$$f_1(t) = f(t), \quad f_2(t) = f_1(-t) = \overline{f_1(t)}$$

satisfy (7.4.7).

The general problem of determining all pairs of characteristic functions satisfying (7.4.7) has still not been solved.

We finally mention a class of distributions which have a remarkable property investigated by J. L. Teugels (1971).

Let $p(x)$ be a probability density with characteristic function $f(t)$. Teugels studied the family of density functions which have the property that

(7.4.8) $f(t) = ap(bt)$

for some positive constants a and b.

Since $p(x)$ is real and integrable one sees that the same is true for $f(t)$, i.e., $f(t)$ corresponds to a symmetric distribution; hence

$$p(x) = p(-x) \quad \text{and} \quad f(t) = f(-t).$$

The inversion formula yields

$$p(x) = \frac{1}{2\pi} \int_{-\infty}^{\infty} e^{-itx} f(t)\,dt$$

and in view of (7.4.8) one gets

$$p(0) = \frac{a}{2\pi b}.$$

Since, according to (7.4.8), $f(0) = ap(0)$ while $f(0) = 1$, we see that

$$a^2 = 2\pi b.$$

Using this relation and the symmetry of $p(x)$ we conclude that $p(x)$ must satisfy the equation

(7.4.9) $$p(x) = \sqrt{\frac{2}{\pi b}} \int_0^\infty \cos\frac{xt}{b}\, p(t)\,dt,$$

where

$$p(0) = (2\pi b)^{-1/2}.$$

Let $p_b(x)$ be a solution of (7.4.9) which corresponds to b; then for any $c > 0$

$$p_c(x) = \sqrt{\frac{b}{c}}\, p_b\!\left(x\sqrt{\frac{b}{c}}\right)$$

and one can assume that $b = 1$.

We say that a frequency function $p(x)$ belongs to the class $\mathscr{P}$ if

(7.4.10a) $$p(x) = \sqrt{\frac{2}{\pi}} \int_0^\infty \cos xt\, p(t)\,dt$$

(7.4.10b) $p(0) = (2\pi)^{-1/2}.$

Theorem 7.4.8. If $p \in \mathscr{P}$ then p is integrable and bounded and $p(x)$ cannot belong to a finite distribution.

It follows from (7.4.10a) that

$$p(x) \leq \sqrt{\frac{2}{\pi}} \int_0^\infty p(t)\, dt \leq \frac{1}{\sqrt{2\pi}},$$

so that $p(x)$ is bounded.

We give an indirect proof for the statement that the frequency $p(x)$ cannot belong to a finite distribution and assume therefore that $p(x) = 0$ for $|x| > x_0 > 0$. The corresponding characteristic function $f(t)$ is therefore an entire function. By our assumptions it coincides with $p(x) = 0$ for $\operatorname{Re} x \geq x_0$ and is therefore identically zero. This is in contradiction with $f(0) = 1$, so that the statement is proved.

Teugels (1971) then uses integral transforms to study the properties of the class $\mathscr{P}$ and he also constructs functions belonging to $\mathscr{P}$.

S. M. Shah (1976) has studied entire functions whose Fourier transforms vanish outside a finite interval, and he obtained as a consequence results concerning certain characteristic functions. We first give two definitions.

The entire function $f(z)$ has bounded index if there exists an integer N such that for all z and all k

$$\max_{0 \leq j \leq N} \left\{ \frac{|f^{(j)}(z)|}{j!} \right\} \geq \frac{|f^{(k)}(z)|}{k!}.$$

The least such integer N is called the index of $f(z)$.

Let $N_R(w, f, z_0)$ denote the number of zeros of $f(z) - w$ in the circle $|z - z_0| < R$. If there exists a constant $c(R)$ such that $N_R(w, f, z_0) < c(R)$ for varying z_0 and w, then $f(z)$ is said to have bounded value distribution.

Theorem 7.4.9. Suppose that the function $p(x)$ is absolutely continuous and non-negative on the closed interval $[a, b]$ and that

$$\int_a^b p(x)\, dx = 1 \quad \text{and} \quad p(a) \neq 0,\ p(b) \neq 0.$$

Then the characteristic function $f(t) = \int_a^b e^{itx} p(x)\, dx$ *is an entire function of bounded index. If* $ab \neq 0$, *then* f *and all successive derivatives* $f^{(k)}$ *have bounded index and bounded value distribution.*

An example is the rectangular distribution over $(a - r, a + r)$.

In a subsequent paper, Shah (1977) studies exceptional values of entire characteristic functions. Some results which are extensions of Marcinkiewicz's theorem are derived.

We have already considered [see Section 4.2 above] the class of self-decomposable distributions. Another important class is the class I_0 of infinitely divisible distributions which have no indecomposable factors.

We next define an additional and interesting family of infinitely divisible characteristic functions [distribution functions] which is called the $\mathscr{L}$-class.

Characteristic functions of the $\mathscr{L}$-class are defined by the following properties:

(i) The Poisson spectrum of a characteristic function $f \in \mathscr{L}$ is either finite or denumerable. Therefore $f(t)$ admits the representation

$$\log f(t) = iat - \gamma t^2 + \sum_{r=1}^{2} \sum_{m=-\infty}^{+\infty} \lambda_{m,r} \left[e^{i\mu_{m,r}t} - 1 - \frac{i\mu_{m,r}t}{1 + \mu_{m,r}^2} \right]$$

where a is real, $\gamma \geq 0$, $\lambda_{mr} \geq 0$ $(r = 1, 2; m = 0, \pm 1, \pm 2, \ldots)$, $\mu_{m,1} > 0$, $\mu_{m,2} < 0$;

(ii) $\sum_{r=1}^{2} \sum_{m=-\infty}^{+\infty} \lambda_{m,r} \mu_{m,r}^2 (1 + \mu_{m,r}^2)^{-1} < \infty$;

(iii) $\sum_{|\mu_{m,r}| < \epsilon} \lambda_{m,r} \mu_{m,r}^2 \to 0 \quad \text{as} \quad \epsilon \to 0$;

(iv) The quotients $\mu_{m+1,r}/\mu_{m,r}$ $(r = 1, 2; m = 0, \pm 1, \ldots)$ are natural numbers greater than 1.

Some theorems concerning the $\mathscr{L}$-class were discussed in Lukacs (1970). We bring here some more recently obtained results concerning this class.

It is known that an infinitely divisible distribution with a Gaussian component which belongs to I_0 belongs also to the $\mathscr{L}$-class. The presence of a Gaussian component is essential, since there exist infinitely divisible distributions without a Gaussian component which are in I_0 but not in $\mathscr{L}$ [see Yu. V. Linnik & I. V. Ostrovskii (1977)].

A. A. Goldberg and I. V. Ostrovskii (1967) constructed an example which shows that there exist characteristic functions which belong to the class $\mathscr{L}$ but not to I_0. This example is the characteristic function

$$f(t) = \exp\left\{ \sum_{k=0}^{\infty} e^{-2^k}(e^{it2^k} - 1) \right\}.$$

It is obvious that $f(t) \in \mathscr{L}$. To show that $f(t)$ is not in I_0 we put

$$f_1(t) = \exp\{a \, e^{-3}(e^{3it} - 1)\} \qquad [0 < a < (\tfrac{2}{3})^3]$$

$$f_2(t) = \exp\left\{ \sum_{k=0}^{\infty} e^{-2^k}(e^{it2^k} - 1) - a \, e^{-3}(e^{3it} - 1) \right\}.$$

Then $f_1(t)$ is a characteristic function and $f(t) = f_1(t) f_2(t)$. It is possible to show that $f_2(t)$ is a characteristic function which is not infinitely divisible; hence it has an indecomposable factor which is also a factor of $f(t)$, so that $f(t)$ cannot belong to I_0.

The above example of Goldberg & Ostrovskii is also presented in Linnik & Ostrovskii (1977).

A. E. Fryntov & G. P. Čistjakov (1977) have studied lattice distributions and obtained a necessary and sufficient condition which assures that they belong to the class I_0.

For the formulation of their result the following definitions are needed.

Let D be a subset of the set Z of (positive or negative) integers and let p be an integer which is different from zero. If $p > 0$ then D_*, D_p and D^p are the intersections of D with the intervals $(-\infty, 0), (0, p)$ and $(p, +\infty)$ respectively. If $p < 0$ then they are the intersections of D with $(0, +\infty), (p, 0)$ and $(-\infty, p)$. We denote the greatest common divisor of these sets by d_*, d_p, d^p. Let $M^+(D)$, respectively $M(D)$, be the smallest additive group which contains D. We define $M(\phi) = M^+(\phi) = \phi$. ($\phi$ is the empty set.)

We say that the set D satisfies the condition (L) if, for any $p \neq 0$,

$$p \notin M(D^p) \cap M^+(D_p \cup \{d_* \operatorname{sign}(p)\})$$

if $D_* \neq \phi$, but

$$p \notin M(D^p) \cap M^+(D_p)$$

if $D_* = \phi$.

Theorem 7.4.10. *Let $F(x)$ be an infinitely divisible lattice distribution with characteristic function $f(t)$ of the form*

$$f(t) = \exp\{i\beta t + \sum_{k \in Z} \lambda(k) e^{(ik\xi t - 1)}\},$$

where β and $\xi \neq 0$ are real, and suppose that the

$$\limsup_{|k| \to \infty} \frac{|k| \log |k|}{\log (1/\lambda(k))} < \infty.$$

Then F belongs to I_0 if and only if the set $D(\lambda) = \{k: \lambda(k) \neq 0\}$ satisfies the condition (L).

The paper also contains examples of sets which satisfy condition (L).

7.5 Convolutions of Poisson type distributions

The results discussed in this section are not new; however, they were not presented in Lukacs (1970), and since they are very interesting they will be briefly treated here.

In his remarkable papers P. Lévy (1937b, c)[*] studied convolutions of Poisson type distributions. A finite convolution of Poisson type distributions

---

[*] See also P. Lévy (1976), pp. 365–417 and R. Cuppens (1975), Appendix B, pp. 217-33.

has a characteristic function of the form

(7.5.1) $$f(t) = \prod_{j=1}^{p} \exp\left[a_j(e^{it\sigma_j} - 1)\right].$$

If we put $e^{it} = x$ and suppose that $\sigma_j = j$, that is, if we make the transition from the characteristic function $f(t)$ to the generating function $F(x)$, we obtain

(7.5.2a) $F(x) = \exp[P(x)]$,

where $P(x)$ is the polynomial

(7.5.2b) $$P(x) = \sum_{j=0}^{p} a_j x^j \qquad [a_p \neq 0, p \text{ positive integer}].$$

Lévy studied exponentials of polynomials, i.e. functions of the form (7.5.2a), and determined conditions which ensured that an exponential of a polynomial be a generating function, that is, that

$$F(x) = \exp[P(x)] = \sum_{k=0}^{\infty} A_k x^k,$$

where all A_k are non-negative. It is here no restriction to assume that $A_0 = 1$. An obvious necessary (but not sufficient) condition for all A_j to be non-negative is that $a_p > 0$, while $a_j \geq 0$ ($j = 1, \ldots, n$) is a trivial sufficient condition. However, all the A_k can be non-negative even if $P(x)$ has some negative coefficients. To formulate Lévy's result we have to introduce the following notations:

(a) Let $P_m(x)$ be the sum of all terms of $P(x)$ whose exponents are at most equal to m.
(b) Let $\bar{P}(x)$ be the sum of all terms in $P(x)$ which have positive coefficients.
(c) Let $\bar{P}_m(x)$ be the sum of all terms of $\bar{P}(x)$ whose exponents are at most equal to m.
(d) Let δ, δ_m and δ'_m ($m < p$) be the greatest common divisors of the degrees of non-zero terms in $P(x)$, $P(x) - P_m(x)$, and $\bar{P}(x) - \bar{P}_m(x)$, respectively.

We now introduce two conditions which will be needed in the sequel.

Condition $\mathcal{A}$. For each integer m for which $a_m < 0$, δ_m exists and is a factor of m. We remark that condition $\mathcal{A}$ depends only on the partition of the coefficients into three groups, namely those which are positive, zero or negative.

Let E_m be the set of all numbers $h_1 p_1 + h_2 p_2 + \ldots + h_v p_v$ where the p_j ($j = 1, \ldots, v$) are the exponents of the non-zero terms in $P_m(x)$, while the h_j ($j = 1, \ldots, v$) are non-negative integers.

Condition $\mathcal{B}$. If the coefficient a_m in $P(x)$ is negative for an m, then m belongs to the set E_m.

Theorem 7.5.1. In order that all A_m should be non-negative, conditions $\mathscr{A}$ and $\mathscr{B}$ are necessary. If these conditions are satisfied and if those a_m which are non-negative are known, then a necessary and sufficient condition for the non-negativity of the A_k can be expressed by means of a finite number of inequalities.

Remark. These conditions are certainly satisfied if the absolute values of the negative coefficients are sufficiently small.

The factorization of exponentials of polynomials, $\exp[P(x)]$, corresponds to the factorization of characteristic functions, obtained by putting $x = e^{it}$.

Lévy also gave a necessary and sufficient condition for the existence of at least one indecomposable component of the characteristic function defined by a polynomial which satisfies conditions $\mathscr{A}$ and $\mathscr{B}$.

7.6 I_0 contains Poisson type distributions

We have seen that the convolution of two Poisson type distributions belongs to I_0, while convolutions of three or more such distributions can have an indecomposable factor, i.e. they may belong to $\bar{I}_0$. Cuppens (1968) gave sufficient conditions which ensure that a finite convolution of Poisson type characteristic functions belongs to I_0. We mention here his result.

Theorem 7.6.1. Let $f(t)$ be the characteristic function

$$f(t) = \exp\left\{\sum_{r=1}^{2}\sum_{j=1}^{p}\sum_{k_j=1}^{n_{r,j}} \lambda_{r,k_j}[\exp(\mathrm{Im}_{r,k_j}\sigma_j t) - 1]\right\}$$

for which the following conditions are satisfied:

(i) $0 \leqslant n_{r,j} < \infty$ (*if $n_{r,j} = 0$ then the corresponding sum is omitted*).
(ii) *the λ_{r,k_j} and the σ_j are real numbers and the $\sigma_1, \ldots, \sigma_p$ are rationally independent.*
(iii) *$m_{1,k_j} \geqslant 0$, $m_{2,k_j} \leqslant 0$ are integers such that $m_{r,k_{j+1}}/m_{r,k_j}$ is an integer greater than 1.*

Then $f(t)$ belongs to I_0.

A. E. Fryntov (1976) studied convolutions of denumerably many Poisson laws of the form

(7.6.1) $\quad f(t) = \exp\left[i\beta t + \sum_\lambda h(\lambda)(e^{it\lambda} - 1)\right].$

Here $h(\lambda) \geqslant 0$ and $\sum_\lambda h(\lambda) < \infty$. We write $D_h = S_h$ for the discrete spectrum of function $h(\lambda)$, that is, D_h is the set of all λ for which h_λ is not zero. Let A be a set of real numbers, then $A^p = A \cap (p, \infty)$ while $A_p = A \cap (-\infty, p)$.

We write $M(A)$ [resp. $M^+(A)$] for the set of all linear combinations with integer coefficients [resp. with positive integer coefficients] of points of A. We quote the main result of Fryntov's paper.

Theorem 7.6.2. *Suppose that the infinitely divisible characteristic function has the form (7.6.1) and that*

(i) *D_h is a subset of the set of positive rational numbers;*
(ii) *there exists a $K > 0$ such that*

$$\sum_{|\lambda| > y} h(\lambda) = O\{\exp(-Ky^2)\} \qquad (y \to \infty).$$

Then $f(t) \in I_0$ if and only if $p \notin M[(D_h)^p]$ for all $p \in (2)M^+(D_h)$. Here $(2)M[(D_h)^p]$ is the vectorial sum $M^+(D_h)(+)M^+(D_h)$.

Remark. In conclusion we note that a great many of the results on the class I_0 deal with analytic characteristic functions. G. P. Čistjakov (1971) showed that there exist in I_0 characteristic functions which are neither analytic nor boundary characteristic functions.

8 Infinitely divisible distributions determined by their values on a half-axis

According to I. A. Ibragimov (1977), Kolmogorov raised the interesting question in the seminar at Moscow University in the 1950s, whether the normal distribution is an infinitely divisible distribution which is determined by its values on a half-axis.

8.1 The case of the normal distribution

This uniqueness property of the normal distribution is formulated in the following theorem.

Theorem 8.1.1. Let $F(x)$ be an infinitely divisible distribution function and let $\Phi(x)$ be the standardized normal distribution

$$\Phi(x) = \frac{1}{\sqrt{2\pi}} \int_{-\infty}^{x} e^{-y^2/2}\, dy.$$

Suppose that $F(x) = \Phi(x)$ for $x < 0$; then $F(x) \equiv \Phi(x)$ for all x.

The proof of the theorem is due to H. J. Rossberg (1974). We follow it here. For the proof we need several lemmas.

Lemma 8.1.1. Let $F(x)$ be an infinitely divisible distribution and suppose that for all $y > 0$

$$\int_{-\infty}^{\infty} e^{-yx}\, dF(x) < \infty.$$

Then

$$f(z) = \int_{-\infty}^{\infty} e^{izx}\, dF(x)$$

is regular at least in the upper half-plane, and we see from the Lévy canonical representation [cf. Theorem 1.3.4] that

$$(8.1.1) \quad \log f(iy) = -ay + \sigma^2 y^2/2 + \int_{-\infty}^{-0} r(y, u)\, dM(u) + \int_{+0}^{\infty} r(y, u)\, dN(u).$$

For convenience we write here

(8.1.1a) $\quad r(y,u) = e^{-yu} - 1 + \dfrac{yu}{1+u^2}.$

The statement of the lemma is plausible and we do not bring its proof here.

Lemma 8.1.2. *Under the assumption of Lemma 8.1.1 the function $f(iy)$ increases at least exponentially if $M(u) \not\equiv 0$.*

Since

$$r(y,u) = e^{-yu} - 1 + yu - y\frac{u^3}{1+u^2}$$

one sees that for $\eta > 0$

(8.1.2) $\quad \displaystyle\int_{+0}^{\infty} r(y,u)\, dN(u)$

$$= y \int_{+0}^{\eta} k(yu)\, u\, dN(u) - y \int_{+0}^{\infty} \frac{u^3}{1+u^2}\, dN(u) + \int_{\eta}^{\infty} r(y,u)\, dN(u),$$

where

(8.1.3) $\quad k(v) = v^{-1}(e^{-v} - 1 + v) > 0.$

Therefore

$$\int_{+0}^{\infty} r(y,u)\, dN(u) \geq -y \int_{0}^{\eta} \frac{u^3}{1+u^2}\, dN(u) - \int_{\eta}^{\infty} dN(u) = K_{\eta}(y) \quad \text{(say)}.$$

We select two real numbers p and q so that $-\infty < p < q < 0$ and assume that $M(q) > M(p)$. We obtain from (8.1.1) the inequality

$$\log f(iy) \geq -ay + \sigma^2 y^2/2 + \int_{r}^{q}\left(e^{y|u|} - 1 - \frac{y|u|}{1+u^2}\right) dM(u) - K_{\eta}(y),$$

which proves the assertion of Lemma 8.1.2.

Lemma 8.1.3. *Let*

$$H(x) = \begin{cases} 2\Phi(x) & \text{if } x < 0 \\ 1 & \text{if } x > 0. \end{cases}$$

The characteristic function of $H(x)$ is then

(8.1.4) $\quad h(t) = e^{-t^2/2}\left[1 - i\sqrt{2/\pi} \int_{0}^{t} e^{w^2/2}\, dw\right],$

INFINITELY DIVISIBLE DISTRIBUTIONS 107

so that

(8.1.5) $\quad h(iy) = e^{y^2/2} [2 + o(e^{-y^2/2})] \quad$ as $y \to \infty$.

We have

$$f(t) = \sqrt{2/\pi} \int_{-\infty}^{0} e^{itu} e^{-u^2/2} \, du = \sqrt{2/\pi} \, e^{-t^2/2} \int_{-\infty}^{0} e^{-(it-u)^2/2} \, du.$$

We evaluate the last integral by contour integration and see that

(8.1.6) $\quad 0 = \int_{Z}^{0} e^{-(it-\zeta)^2/2} \, d\zeta + \left[\int_{0}^{it} + \int_{it}^{it+Z} + \int_{it+Z}^{Z} (e^{-(it-\zeta)^2/2} \, d\zeta) \right].$

This leads to (8.1.4) which in turn implies (8.1.5).

Lemma 8.1.4. *Let $f_1(t)$ be the characteristic function of an infinitely divisible distribution function $F_1(x)$. Then $F_1(0) = 0$, $F_1(\epsilon) > 0$ if and only if in its Lévy canonical representation [determined by $\alpha_1, \sigma_1, M_1, N_1$] one has $\sigma_1^2 = 0$ and $M_1(u) \equiv 0$ for $u < 0$, while*

$$\int_{0}^{1} u \, dN(u) < \infty$$

and

$$a_1 - \int_{+0}^{\infty} \frac{u}{1+u^2} \, dN_1(u) = 0.$$

Then

$$\log f_1(iy) = \int_{+0}^{\infty} (e^{-yu} - 1) \, dN_1(u).$$

The proof of this lemma would not fit the topics discussed in this chapter; moreover it is a consequence of Theorem 11.2.2 of Lukacs (1970). We proceed therefore to the proof of Theorem 8.1.1.

Let

$$G(x) = \begin{cases} 2F(x) - 1 & \text{if } x > 0 \\ 0 & \text{if } x < 0. \end{cases}$$

Then $F(x)$ has the representation

$$F(x) = \tfrac{1}{2} [H(x) + G(x)].$$

Let $g(t)$ be the characteristic function of $G(x)$. It follows from Lemma 8.1.4

that

$$f(iy) = \int_{-\infty}^{\infty} e^{-yx}\, dF(x) = \tfrac{1}{2}[h(iy) + g(iy)]$$
$$= e^{y^2/2}[1 + O(e^{-y^2/2})].$$

Therefore

(8.1.7) $\quad \log f(iy) = y^2/2 + O(e^{-y^2/2}).$

We compare the asymptotic behaviour of (8.1.7) and (8.1.1) as $y \to \infty$ and conclude from Lemma 8.1.1 that $M(u) \equiv 0$ for $u < 0$.

According to Theorem 1.3.4,

$$\int_{+0}^{1} u^2 \, dN(u)$$

is finite. We have to consider two possibilities:

(I) $\quad \displaystyle\int_{+0}^{\infty} u \, dN(u) = \infty.$

The function $k(u)$, introduced in (8.1.3), is increasing and positive. Let

$$J(u) = \int_{1}^{u} v \, dN(v).$$

Then $J(+0) = -\infty$ and we define

$$R_\eta(y) = \int_0^\eta k(yu)\, u\, dN(u) > \int_1^{y\eta} k(v)\, dJ\!\left(\frac{v}{y}\right) \geq e^{-1}[J(\eta) - J(y^{-1})].$$

Therefore we see from (I) that

$$\lim_{y \to \infty} R_\eta(y) = \infty.$$

It follows from (8.1.1) and (8.1.2) that

$$\log f(iy) = \sigma^2 y^2/2 + y\left[-a + R_\eta(y) - \int_0^\eta \frac{u^3}{1+u^2}\, dN(u)\right.$$
$$\left. + \int_\eta^\infty \frac{u}{1+u^2}\, dN(u)\right] + \int_\eta^\infty (e^{-yu} - 1)\, dN(u),$$

so that

$$\log f(iy) = \sigma^2 y^2/2 + y R_\eta(y)[1 + o(1)] \qquad \text{as } y \to \infty.$$

Using (8.1.7) we obtain the relation

(8.1.8) $\quad y^2/2 + O(e^{-y^2/2}) = \sigma^2 y^2/2 + y R_\eta(y)[1 + o(1)]$

INFINITELY DIVISIBLE DISTRIBUTIONS

as $y \to \infty$.

We again distinguish two cases:

(Ia) $\quad \lim_{y \to \infty} y^{-1} R_\eta(y) = 0$.

Then $\sigma = 1$ and

(8.1.9) $\quad yR_\eta(y) = O(e^{-y^2/2})$,

which is a contradiction. If

(Ib) $\quad \lim_{y \to \infty} y^{-1} R_\eta(y) = l_\eta > 0$,

then

$$\sigma^2/2 + l_\eta = \tfrac{1}{2}.$$

Since $v^{-1} k(v)$ is bounded, say $v^{-1} k(v) < c$ for all $v > 0$, we have

$$yR_\eta(y) = y^2 \int_{+0}^{\eta} \left[\frac{k(yu)}{yu} \right] u^2 \, dN(u) < cy^2 \int_{+0}^{\eta} u^2 \, dN(u).$$

Hence l_η can be made arbitrarily small. Then we again obtain $\sigma = 1$ and we see from (8.1.8) that (8.1.9) holds. This is also a contradiction, so that (I) is false.

We now assume

(II) $\quad \int_{+0}^{1} u \, dN(u) < \infty$.

Then we obtain from (8.1.1) and (8.1.1a), after a simple computation,

(8.1.10) $\quad \log f(iy) = \sigma^2 y^2/2 + y \left[-a + \int_{+0}^{\infty} \frac{u}{1+u^2} \, dN(u) \right]$

$$+ S_\eta(y) + \int_\eta^\infty (e^{-yu} - 1) \, dN(u),$$

where

$$S_\eta(y) = \int_{+0}^{\eta} (e^{-yu} - 1) \, dN(u) = y \int_{+0}^{\eta} \frac{e^{-yu} - 1}{yu} \, d\int_{+0}^{u} v \, dN(v).$$

It is easily seen that

(8.1.11) $\quad -y \int_{+0}^{\eta} v \, dN(v) \leqslant S_\eta(y) \leqslant 0$.

We compare (8.1.10) and (8.1.8) and see that $\sigma = 1$ and that

(8.1.12) $\quad \left[-a + \int_{+0}^{\infty} \frac{u}{1+u^2} \, dN(u) \right] + \frac{1}{y} S_\eta(y) = o(1)$.

We write
$$\lim_{y \to \infty} y^{-1} S_\eta(y) = s_\eta$$
and get from (8.1.12)

(8.1.13) $\quad -a + \int_{+0}^{\infty} \dfrac{u}{1+u^2} \, dN(u) + s_\eta = 0.$

It follows from (8.1.11) that
$$|s_\eta| \leq \int_{+0}^{\eta} v \, dN(v),$$
so that $|s_\eta|$ can be made arbitrarily small and
$$-a + \int_{+0}^{\infty} \dfrac{u}{1+u^2} \, dN(u) = 0.$$

Lemma 8.1.3 and formula (8.1.10) imply that

(8.1.14) $\quad f(iy) = e^{\sigma^2 y^2 / 2} f_1(iy).$

Here $f_1(t)$ is the characteristic function of $F_1(x)$, where $F_1(0) = 0, F_1(\epsilon) > 0$. On account of (8.1.7), equation (8.1.14) is only possible if $f_1(y) \equiv 1$, that is, if $N_1(u) = N(u) \equiv$ constant. We then see from (8.1.13) that $a = 0$ and the proof is completed.

M. Riedel (1975) has sharpened somewhat Rossberg's result.

8.2 Generalizations of Theorem 8.1.1

I. A. Ibragimov (1977) has generalized Theorem 8.1.1 considerably and he obtained the following result.

Theorem 8.2.1. Let $f(t)$ be the characteristic function of an infinitely divisible distribution $F(x)$ and suppose that $f(t)$ can be continued analytically into the upper (lower) half-plane of the complex variable $z = t + iy$. Assume further that the infinitely divisible distribution $G(x)$ coincides with $F(x)$ on the half-line $[-\infty, a)$ (respectively on the half-line $[a, \infty)$). Then $G(x)$ is either zero (respectively one) on some half-line or $F(x) = G(x)$ for all x.

The last two theorems stimulated additional work concerning the analytic continuation of infinitely divisible distribution functions which are given on a suitable subset of the real axis. Here we mention B. Jesiak (1979) and G. Siegel (1979) who have studied the analytic continuation of symmetric distribution functions. In these studies the theory of analytic distribution functions is an essential tool. A survey of the continuation theory of analytic distribution functions was given by Rossberg, Jesiak & Siegel (1981).

INFINITELY DIVISIBLE DISTRIBUTIONS

The next theorem requires assumptions concerning the tail behaviour of the distribution function. Some related results are also given in Steutel (1974).

Theorem 8.2.2. Let $f(t)$ be an infinitely divisible characteristic function and suppose that there exist constants a and $\alpha (a > 0, 0 < \alpha \leq 1)$ such that $T(x) = O[\exp(-ax^{1+\alpha})]$; then $f(t)$ is the characteristic function of a (possibly degenerate) normal distribution.

The function $f(t)$ is an entire function of order $\rho \leq 1 + \alpha^{-1}$. Since f is infinitely divisible it has no zeros, so that $f(z) = \exp[g(z)]$. It follows from the definition of ρ that

$$\max_{|z|=r} \operatorname{Re} g(z) = \max_{|z|=r} \log |f(z)| = \log \max_{|z|=r} |f(z)| \leq r^{\rho + \epsilon}$$

for any $\epsilon > 0$ and all sufficiently larger r. We therefore conclude [cf. R. P. Boas (1954), theorem 1.3.4] that $g(z)$ is a polynomial of order not exceeding ρ. From the theorem of Marcinkiewicz we see that $\rho \leq 2$, so that Theorem 8.2.1 is proved.

9 A generalization of the decomposition problem

In this chapter we discuss an analytical extension of the decomposition of characteristic functions.

9.1 α-decompositions

A characteristic function $f(t)$ is said to admit a finite α-decomposition if it satisfies a relation of the form

$$(9.1.1) \quad f(t) = \prod_{j=1}^{n} [f_j(t)]^{\alpha_j},$$

either in an interval $(-\delta, +\delta)$ in which $f(t)$ does not vanish, or in a sequence $\{t_k\}$ of points such that $\lim_{k \to \infty} t_k = 0$. Here $\alpha_1, \ldots, \alpha_n$ are positive numbers, while the $f_j(t)$ are characteristic functions.

In a similar way one defines denumerable α-decompositions by a relation

$$(9.1.1\text{a}) \quad f(t) = \prod_{j=1}^{\infty} [f_j(t)]^{\alpha_j}.$$

Denumerable α-decompositions are possible. This is indicated by the following theorem.

Theorem 9.1.1. *Let $f_j(t)$ be a sequence of characteristic functions and suppose that $0 < \alpha_1 < \alpha_2 < \ldots < \alpha_n < \ldots$ is a sequence of positive numbers. Suppose that in a neighbourhood of the origin*

(i) $\quad \prod_{j=1}^{\infty} [f_j(t)]^{\alpha_j} = f(t),$

where $f(t)$ is an analytic characteristic function which has no zeros in its strip of regularity. Then the functions $f_j(t)$ are analytic characteristic functions and are regular at least in the strip of regularity of $f(t)$, and relation (i) is valid in this strip.

Here it is assumed that relation (i) holds either in an interval or on a sequence of points which tends to zero. The $\alpha_j > 0$ and the $f_j(t)$ are characteristic functions. The $f_j(t)$ are called the α-components (or α-factors) of $f(t)$. For the proof we refer the reader to Lukacs (1970).

Remark. An indecomposable distribution may admit an α-decomposition.

This is shown by the following example. We consider the characteristic functions

$$f(t) = \tfrac{1}{35}[1 + 2\,e^{it} + 5\,e^{2it} + 12\,e^{3it} + 15\,e^{4it}]$$

$$f_1(t) = \tfrac{1}{7}[1 + 3\,e^{it} + 3\,e^{2it}]$$

$$f_2(t) = \tfrac{1}{175}[1 + e^{it} + 8\,e^{2it} + 17\,e^{3it} + 28\,e^{4it} + 45\,e^{5it} + 75\,e^{6it}].$$

It is easily seen that $f(t)$ is indecomposable and that

$$[f(t)]^2 = f_1(t)\,f_2(t).$$

Hence the indecomposable characteristic function admits the α-decomposition

$$f(t) = [f_1(t)]^{1/2}\,[f_2(t)]^{1/2}.$$

It is known that α-factors of a normal distribution are necessarily normal distributions [Theorem 10.2.2 in Lukacs (1970)]. This admits the following generalization.

Theorem 9.1.2. Suppose that the functions $f_1(t)$, $f_2(t)$,..., $f_s(t)$ are characteristic functions which satisfy the relation

$$(9.1.2) \quad \prod_{j=1}^{s}[f_j(t)]^{\alpha_j} = \exp[Q(t)].$$

Here $Q(t)$ is a polynomial of degree n such that $Q(-t) = \overline{Q(t)}$, while the α_j are positive real numbers. Then $n \leqslant 2$.

For the proof we refer to Yu. V. Linnik & I. V. Ostrovskii (1977).

D. Dugué raised the following question. Suppose that the characteristic function $f(t)$ is analytic in a half-plane (say, in $\operatorname{Im} z < 0$) and assume that $f(t)$ admits an α-decomposition. Are then the α-components of $f(t)$ analytic in $z < 0$?

G. P. Čistjakov (1975) solved this problem by proving the following theorem.

Theorem 9.1.3. Suppose that the characteristic function $f(t)$ is analytic in the semicircle $\{|z| < R,\ \operatorname{Im} z > 0\}$ and is continuous on its closure. If the characteristic functions $f_1(t),\ldots,f_n(t)$ are such that

$$(9.1.3) \quad f(t) = \prod_{j=1}^{n}[f_j(t)]^{\alpha_j}$$

in some interval $|t| < \delta$, $0 < \delta < R$, then the characteristic functions $f_j(t)$, $j = 1,\ldots,n$ are analytic in the semicircle $\{|z| < R,\ \operatorname{Im} z > 0\}$.

Corollary to Theorem 9.1.3. Let $F(x)$ [resp. $F_j(x)$] be the distribution function which corresponds to the characteristic function $f(t)$ [resp. $f_j(t)$] and suppose that equation (9.1.3) holds. If the support of $F(x)$ is bounded to the left (right) then this is also true for all α-components of $F(x)$.

B. Ramachandran (1967) [pp. 155-7] proved the following extensions of Theorem 9.1.1.

Theorem 9.1.4. Suppose that the assumptions of Theorem 9.1.3 are satisfied and that $f(t) = \exp[i\mu t - \sigma^2 t^2/2]$; then the characteristic functions $f_j(t)$ are characteristic functions of normal distributions.

Theorem 9.1.5. Suppose that the assumptions of Theorem 9.1.1 are satisfied and that $f(t) = \exp[\lambda(e^{it} - 1)]$ ($\lambda \geq 0$); then the characteristic functions $f_j(t)$ belong to Poisson distributions.

Denumerable α-decompositions were also studied by Linnik & Ostrovskii (1977), p. 248, by Ramachandran (1967), pp. 152-3, and by R. G. Laha (1969).

9.2 The class I_0^α

Linnik introduced the class I_0^α of infinitely divisible characteristic functions which have only infinitely divisible α-components. This class is analogous to the class I_0; clearly $I_0^\alpha \subset I_0$.

Theorem 9.2.1. Let $f(t)$ be an infinitely divisible characteristic function which belongs to the class $\mathscr{L}$. Suppose that

$$f(t) = [f_1(t)]^{\alpha_1} \ldots [f_s(t)]^{\alpha_s},$$

where the α_j are positive and where the $f_j(t)$ are characteristic functions. Assume further that for some $k > 0$,

$$\lambda_{m,r} = O\left[\exp(-k\mu_{m,r}^2)\right]$$

as $m \to \infty$ and $r = 1, 2$. Then $f(t) \in I_0^\alpha$. The $\lambda_{m,r}$ and the $\mu_{m,r}$ were introduced when we defined the $\mathscr{L}$-class.

In view of the assumptions of the theorem one can conclude from Theorem 9.2.1 [Lukacs (1970)] that $f(t) \in I_0$. It follows from Theorem 9.1.1 (*ibid.*) that $f(t)$ is an entire function without zeros. Then the $f_j(t)$ ($j = 1, \ldots, s$) are also entire functions without zeros and the $[f_j(t)]^{\alpha_j}$ are ridge functions. It can be shown that

$$[f_j(t)]^{\alpha_j} = \exp\left[i\beta_j t + \int_{-\infty}^{\infty} \left(e^{itx} - 1 - \frac{itx}{1+x^2}\right) \frac{1+x^2}{x^2} \, d\theta_j(x)\right],$$

where β_j is a real number and where $\theta_j(x)$ is non-decreasing. This means that the $f_j(t)$ are infinitely divisible characteristic functions with spectral function $\theta_j(x)/\alpha$.

Ostrovskii (1970b) obtained the following sufficient condition which assures that certain infinitely divisible characteristic functions belong to I_0^α.

Theorem 9.2.2. *Let $f(t)$ be an infinitely divisible characteristic function of the form*

$$f(t) = \exp\left\{i\beta t + \int_{-\infty}^{\infty} (e^{itx} - 1) \, dG(x)\right\},$$

where β is real while $G(x)$ is a non-decreasing function which satisfies the following conditions:

(i) *$G(x)$ is a step function whose discontinuity points form a set of rationally independent points;*

(ii) *For a constant $k > 0$,*

$$\int_{|x|>y} dG(x) = O[\exp(-ky^2)] \quad \text{as } y \to \infty.$$

Then $f(t) \in I_0^\alpha$.

Ostrovskii (1970b) also showed that I_0^α is dense in the set of all infinitely divisible distributions in the following sense. Let F be an infinitely divisible distribution; then there exists a sequence $\{F_n\}$ of distributions in I_0^α such that

$$\lim_{n \to \infty} F_n = F.$$

In proving this result Ostrovskii uses the theory of almost periodic functions.

Fryntov (1974) obtained the following result.

Theorem 9.2.3. *Suppose that the characteristic function $f(t)$ has the form*

$$f(t) = \exp\left\{i\beta t + \int_{-\infty}^{\infty} (e^{itx} - 1) \, dG(x)\right\},$$

where the following conditions are satisfied:

(i) *β is real while $G(x)$ is a measure which is concentrated on the set $[a, 2a] \cup \{\lambda_m\}_{m=1}^{\infty}$, where $a > 0$ and the $\lambda_m \geq a$ are such that the ratio λ_{m+1}/λ_m is a positive integer different from 1.*

(ii) *For some $k > 0$,*

$$\int_{|x|>y} dG(x) = O[\exp(-ky^2)] \quad \text{as } y \to \infty.$$

Then $f(t) \in I_0^\alpha$.

10 Boundary characteristic functions

A characteristic function $f(t)$ is called a boundary characteristic function if there exists a complex-valued function $A(z)$ of the complex variable $z = t + iy$ (t, y real) which is regular in the rectangle $R^+ = \{|t| < \Delta, 0 < y < b\}$ [respectively in the rectangle $R^- = \{|t| < \Delta, -a < y < 0\}$] and which has the property that $\lim_{y \to 0} A(t + iy) = f(t)$ in R^+ {respectively in R^-}.

As an example of a distribution which has a boundary characteristic function we mention the stable distribution with parameters $\alpha = \gamma = \frac{1}{2}$. Its density function is

$$p_{1/2, 1/2}(x) = \begin{cases} 0 & \text{for } x < 0 \\ \dfrac{1}{\sqrt{2\pi}} x^{-3/2} e^{-1/(4x)}, & \text{for } x > 0 \end{cases}$$

The corresponding characteristic function is

$$f(t) = \exp\left\{-|t|^{1/2} \exp\left[-\frac{t}{|t|}\frac{\pi}{4}\right]\right\}.$$

The function $A(z) = \int_0^\infty e^{izv} p(v)\, dv$ is not regular in a circle around the origin, but it follows from the Cauchy–Riemann equations that it is regular in the upper half-plane. Hence $f(t)$ is a boundary characteristic function.

In Section 7.1 we described the natural domain of analyticity of an analytic characteristic function. We discuss here the same problem for boundary characteristic functions $f(t)$.

Let $A(z)$ be a complex-valued function such that $A(z)$ is regular in the rectangle $[|t| < \Delta, a < -y < 0]$, and suppose that

$$\lim_{y \uparrow 0} A(t + iy) = f(t).$$

It follows from a theorem due to J. Marcinkiewicz (1938) that $A(z)$ is regular, at least in the strip $\{-a < y < 0, -\infty < t < \infty\}$.

Theorem 10.1.1. *Let G be a domain which satisfies the following conditions:*

(i) *G contains a strip* $-\alpha < \operatorname{Im} z < 0$.
(ii) *G is symmetric with respect to the imaginary axis.*

(iii) *The points $-i\alpha$ and 0 belong to the boundary ∂G of G.*
(iv) *If $b \in CG$ and $\operatorname{Im} b > 0$ then $\overline{\operatorname{Im} b} \in CG$.*
(v) *Every point α of the real axis which belongs to ∂G is the limit of a sequence $x_n + iy \in CG$ such that $y_n > 0$.*

Here CG denotes the complement of G. Then there exists an analytic function $A(z)$ whose domain of analyticity is G and a characteristic function $f(t)$ such that

$$\lim_{y \uparrow 0} A(t + iy) = f(t) \qquad (t \text{ real}).$$

We suppose that G is not the strip $-\alpha < \operatorname{Im} z < 0$ and we can assume, without loss of generality, that $\alpha = -\infty$.

Let $\{b_k\}$ be the sequence constructed according to Lemma 7.1.3 and let $\{\epsilon_k\}$ be a decreasing sequence of positive numbers such that $\epsilon_0 = +\infty$, $\lim_{k \to \infty} \epsilon_k = 0$, $\epsilon_n \neq \operatorname{Im} b_k$ for all k and n. We introduce the sets

$$G_n' = \{z \mid z \in CG, \ \epsilon_n \leq \operatorname{Im} z \leq \epsilon_{n-1}\}$$

and

$$G_n = CG_n'.$$

If ϵ_1 is chosen small enough, G_n is connected and satisfies the conditions of Theorem 7.1.1. We write $f_n(t)$ for the characteristic function constructed according to Theorem 7.1.1; then G_n is the natural domain of regularity of $f_n(t)$. Let $\{a_j\}$ be the sequence determined by Lemma 7.1.3 and consider

$$f(t) = \sum_{j=1}^{\infty} a_j f_j(t).$$

Then $f(t)$ is a characteristic function. We define the region

$$H_{\epsilon, R} = \{z \mid z \in G, |z| < R, d(z, \partial G) > \epsilon\}.$$

Here $d(z, \partial G)$ is the distance of the point z from ∂G. It follows from the definition of $f(z)$ and from the existence of constants K and M [of (7.1.6)], where M depends on ϵ and R but not on f and z, that

$$|f_j(z)| < M$$

if $z \in H_{\epsilon, R}$. Therefore $f(z)$ is regular in G.

We still have to show that the points b_k are singular points of $f(z)$. We distinguish two cases. If $\operatorname{Im}(b_k) > 0$ then b_k belongs to exactly one of the sets G_n' and is therefore a singular point of the function $f_n(z)$. Moreover there exists a constant M' such that $|f_j(z)| < M'$ for $j \neq n$ and $|z - b_n|$ sufficiently small.

Hence the sum

$$\sum_{j \neq n} a_j f_j(z)$$

is finite, so that b_k is a singular point of $f(z)$. We consider next the case where $\mathrm{Im}(b_n) = 0$. According to assumption (v) of Theorem 10.1.1 there exists a sequence $\{\beta_m\}$ such that $\beta_m \in CG$, $\mathrm{Im}(\beta_m) > 0$ and

$$\lim_{m \to -\infty} \beta_m = b_k.$$

Since $f(z)$ is not regular in the points β_m one sees that b_k is a singular point of $f(z)$. The statement of the theorem follows in the case where G is not the strip $\{-\alpha < \mathrm{Im}\, z < 0\}$. An example of such a function is the Weierstrass function

$$f(t) = \sum_{k=0}^{\infty} 2^{-(k+1)} \exp(it5^k).$$

The function $f(-t)$ belongs to a probability distribution which is bounded to the right and is therefore a boundary function of a function which is analytic in $\{\mathrm{Im}\, z < 0\}$. Since f is nowhere differentiable on the real axis, any point of the real axis is a singular point of f.

11 Mixtures of distribution functions

This chapter deals with the mixture of distribution functions from the viewpoint of their infinite divisibility.

11.1 Infinite divisibility of mixtures

Several authors [F. W. Steutel (1971, 1973); D. Kelker (1971); J. Keilson & F. W. Steutel (1972, 1974)] have studied conditions which ensure that a mixture is infinitely divisible. We present here some of their results and note that a mixture of infinitely divisible characteristic functions is not necessarily infinitely divisible. As an example we mention the characteristic function

$$f(t) = \tfrac{1}{2}(e^{-t^2} + e^{-2t^2}).$$

This function is an entire characteristic function but is zero for $t^2 = \pi i$ and is therefore, according to Theorem 1.3.1, not infinitely divisible.

We first consider mixtures of exponential distributions. The density of such a mixture has the form

$$(11.1.1) \quad p(x) = \sum_{j=1}^{n} p_j \lambda_j e^{-\lambda_j x}$$

where $p_j \neq 0$,

$$\sum_{j=1}^{n} p_j = 1$$

and $\lambda_j > 0$. Without loss of generality we can assume that

$$(11.1.1a) \quad 0 < \lambda_1 < \lambda_2 \ldots < \lambda_n.$$

We do not assume that all the p_j are positive since $p(x)$ could be non-negative for all x, even if some of the p_j are negative.[*]

The characteristic function of a density function (11.1.1) has the form

$$(11.1.2) \quad f(t) = \sum_{j=1}^{n} p_j \frac{\lambda_j}{\lambda_j - it} = P(t) \Big/ \prod_{j=1}^{n} (\lambda_j - it),$$

---

[*] D. J. Bartholomew (1969) gave a condition which ensures that a mixture (11.1.1) be a probability density.

where $P(t)$ is a polynomial of degree not exceeding $(n-1)$; thus $f(t)$ can have at most $(n-1)$ zeros.

Theorem 11.1.1. A characteristic function of the form (11.1.2) *which satisfies* (11.1.1a) *is infinitely divisible if the sequence* $p_1, p_2, \ldots, p_n$ *has at most one change of sign.*

If one puts $t = -iv$ (v real) then one sees from (11.1.2) that $f(-iv)$ has at least one zero between every λ_j and λ_{j+1}, provided that p_j and p_{j+1} have the same sign. If all p_j have the same sign then one gets all $(n-1)$ zeros. If there is one change of sign in the sequence $p_1, \ldots, p_n$ then $(n-2)$ zeros of $P(t)$ are obtained. If in this case $\Sigma p_j \lambda_j = 0$ then $P(t)$ is of degree $n-2$ and no more zeros exist. If $\Sigma p_j \lambda_j > 0$ then $f(-iv)$ is negative for large positive values of v, so that a zero is located on the interval (λ_n, ∞).

Thus we see that $P(t)$ has only purely imaginary zeros $t_k = i\mu_k$ (μ_k real, positive). Then $f(t)$ can be written in the form

$$(11.1.3) \quad f(t) = \prod_{j=1}^{n} \frac{\lambda_j}{\lambda_j - it} \prod_{k=1}^{m} \frac{\mu_k - it}{\mu_k},$$

where m is either $(n-1)$ or $(n-2)$ and where $\mu_k > \lambda_k$ ($k = 1, 2, \ldots, m$). To show that $f(t)$ is infinitely divisible we use a particular case of the Lévy-Khinchine canonical representation (Theorem 1.3.4A) by assuming that $\theta(x)$ is absolutely continuous and by writing

$$g(x) = \frac{1+x^2}{x^2} \theta'(x).$$

A characteristic function $f(t)$ whose logarithm can be written in the form

$$(11.1.4) \quad \log f(t) = ita + \int_0^{\infty} \left[e^{itx} - 1 - \frac{itx}{1+x^2} \right] g(x) \, dx$$

is infinitely divisible if a is real, $g(x) \geq 0$ and if $\frac{x^2}{1+x^2} g(x)$ is integrable over $(0, \infty)$. We apply this to the characteristic function $\frac{\lambda}{\lambda - it}$ of an exponential distribution and see that $g(x) = e^{-\lambda x}/x$ while

$$a = \int_0^{\infty} \frac{e^{-\lambda x}}{1+x^2} \, dx.$$

It follows from (11.1.3) that $f(t)$ admits a representation (11.1.4) with

$$xg(x) = \sum_{j=1}^{n} e^{-\lambda_j x} - \sum_{j=1}^{m} e^{-\mu_k x},$$

so that $f(t)$ is infinitely divisible and Theorem 11.1.1 is proved.

MIXTURES OF DISTRIBUTION FUNCTIONS   121

Theorem 11.1.1 is due to F. W. Steutel who considered various generalizations in his monograph (1970) and studied mixtures of Gamma distributions. He also investigated the infinite divisibility of mixtures of the form

$$\left[\frac{\lambda}{\lambda - h(t)}\right]^\alpha \quad (\lambda > 0, \alpha > 0)$$

for suitable functions $h(t)$.

We next give a result on mixtures of geometric distributions. This is similar to Theorem 11.1.1 and was communicated to the author by H. J. Rossberg & G. Siegel. The characteristic function of the geometric distribution is

$$f(t) = \sum_{j=0}^{\infty} q(1-q)^j e^{itj} \quad (0 < q < 1).$$

Theorem 11.1.2. Suppose that the sequence $p_1, p_2, \ldots, p_n$ has at most one change of sign and that $p_1 + p_2 + \ldots + p_n = 1$. Let q_j be real numbers such that $0 < q_1 < \ldots < q_n < 1$. If

$$f(t) = \sum_{j=1}^{n} p_j f_{q_j}(t)$$

is a characteristic function, then it belongs to an infinitely divisible distribution.

The proof is similar to the proof of the preceding theorem.

A survey of mixtures of infinitely divisible distributions can be found in the monograph of K. van Harn (1978) who studied mixtures of infinitely divisible lattice distributions.

11.2 Mixtures of exponential distributions which are not infinitely divisible

There are frequency functions which are mixtures of exponential frequencies and which are not infinitely divisible. An example of such a frequency function was given by Steutel (1967). We take $\lambda_1 = 1, \lambda_2 = 3, \lambda_3 = 5, p_1 = 2, p_2 = -2, p_3 = 1$ and get, according to (11.1.1)

$$p(x) = 2e^{-2x} - 6e^{-3x} + 5e^{-5x}.$$

The function $p(x)$ is non-negative for all x and is therefore a frequency function. The corresponding characteristic function is

$$f(t) = \frac{(15 - t^2)}{(1 - it)(3 - it)(5 - it)}.$$

This function has real zeros and is therefore not infinitely divisible.

Steutel (1967) also gave an example of an infinitely divisible mixture where the $\{p_j\}$ have two changes of sign. This shows that the condition of Theorem 11.1.1 is sufficient but not necessary.

Steutel (1970, 1973) has given an extensive discussion of infinitely divisible mixtures and has treated the infinite divisibility of mixtures of Gamma distributions.

We next consider variance mixture of normal distributions. A distribution $F(x)$ is said to be a variance mixture of normal distributions if its characteristic function $f(t)$ is

(11.2.1) $$f(t) = \int_0^\infty e^{-t^2 y/2} \, dH(y),$$

where $H(y)$ is a non-degenerate distribution function such that lext $H(y) = 0$.

The class of infinitely divisible variance mixtures contains the Laplace distribution, the symmetric stable distributions and also Student's distribution [see Steutel (1968)].

Feller [(1971), p. 538] showed that $f(t)$ is infinitely divisible if $H(y)$ is infinitely divisible. Kelker (1971) constructed a distribution function $H(y)$ which is not infinitely divisible but is such that the variance mixture (11.2.1) is infinitely divisible.

Theorem 11.2.1. *If the mixing distribution $H(y)$ is finite, then the variance mixture* (11.2.1) *cannot be infinitely divisible.*

According to Theorem 7.2.3 the characteristic function $h(t)$ of $H(y)$ has infinitely many zeros in the complex plane. Since $f(t) = h(it^2/2)$ one sees that $f(t)$ is an entire characteristic function with infinitely many zeros and therefore cannot be infinitely divisible.

12 Analytic distribution functions

In Chapter 7 we discussed analytic characteristic functions. The importance of these results suggests similar investigations concerning the analytic properties of distribution functions. In this chapter we present briefly some results of these studies.

12.1 Definition of analytic distribution functions and some of their properties

In the following we denote by H_+ the upper half-plane, $H_+ = \{z: \text{Im}(z) > 0\}$, and by D the real axis.

A distribution function $F(x)$ is said to be an analytic distribution function if there exists a function $R(z)$ of the complex variable $z = x + iy$ such that

(i) $R(z)$ is analytic in a simply connected domain $A \subseteq H_+$ such that the boundary ∂A of A contains D.
(ii) $R(x)$ is continuous and $R(x) = F(x)$ for $x \in D$.

In view of Schwarz' reflection principle we see that an analytic distribution function is analytic in the domain $A \cup A' \cup D$, where $A' = \{z: \bar{z} \in A\}$. From now on we shall write for the analytic continuation $R(z)$ of $F(x)$ also $F(z)$.

One sees that $F(\bar{z}) = \overline{F(z)}$, so that the zeros and the singularities of $F(z)$ are located symmetrically with respect to the real axis.

Theorem 12.1.1. *Let $f(t)$ be the characteristic function of the distribution function $F(x)$. Suppose that for some $c > 0$,*

(12.1.1) $\quad f(t) = O[\exp(-c|t|)] \quad$ as $t \to \infty$.

Then $F(x)$ is analytic in D. More precisely, there exists a function $R(z)$ of the complex variable $z = x + iy$ such that $R(z)$ is analytic in the strip $|y| < c$ and coincides with $F(x)$ on D.

Proof. Let

(12.1.2) $\quad F_n(z) = \dfrac{1}{2\pi} \displaystyle\int_{-n}^{n} \dfrac{1 - e^{-itz}}{it} f(t) \, dt$.

This is an entire function. Consider the strip $|y| < c'$, where $c' < c$ and let

$0 < A < B$. Then

$$\left| \int_A^B \frac{1-e^{-itz}}{it} f(t) \, dt \right| \leq A^{-1} \int_A^B (1 + e^{t|y|}) |f(t)| \, dt$$

$$= \frac{M}{A} \int_A^B (i + e^{tc'}) e^{-ct} \, dt.$$

Here M is a constant such that $|f(t)| \leq M e^{-ct}$; the existence of M follows from (12.1.1). The last integral tends to 0 as A and B go to ∞. Therefore $F_n(z)$ converges uniformly as $n \to \infty$ in $x \in D$, $|y| < c'$ to a limit $R(z)$ which is analytic in the strip $|y| < c$ and which agrees with the given distribution function $F(x)$ on the real axis.

Some early work on analytic distribution functions was done by D. A. Raikov (1939) and by R. P. Boas & F. Smithies (1938).

Raikov considered distribution functions $F(z)$ which are regular in a strip $I_R = |\text{Im}(z)| < R$, $R > 0$. If $F(z)$ is bounded in every strip $|\text{Im}(z)| \leq r < R$ which is located in the interior of I_R, then the convolution of F with any distribution G is regular in I_R and is bounded in every interior strip of I_R. Raikov showed that there exist analytic distribution functions which are not bounded in an arbitrarily narrow strip $|\text{Im}(z)| < \rho$. As an example he mentions the distribution function $F(z) = \exp(1 - e^{e^{-z}})$. An elementary computation shows that $F(z)$ is unbounded if $x \to -\infty$ while $|y| < -\pi/2$. Raikov also showed that there exists an entire distribution function $F(z)$ such that $F(z) * [1 - F(-z)]$ is not regular at $z = 0$. This contrasts with the behaviour of analytic characteristic functions.

Boas & Smithies obtained among other results the following. Let $g(y)$ be an even function which satisfies the following conditions: (i) $g(y) > 0$ for all y; (ii) $g(y) \downarrow 0$ as $y \to \infty$; (iii) $yg(y) \uparrow \infty$ as $y \to \infty$ in $y \geq y_0 > 0$; (iv) $\exp[-yg(y)]$ is convex for $y \geq y_0$. Then there exists a distribution function $F(x)$ which is not analytic in the point $x = 0$ and whose Fourier-Stieltjes transform $f(y)$ satisfies the relation $f(y) = O[\exp(-|y|g(y))]$ as $|y| \to \infty$.

The following statement extends Theorem 12.1.1 essentially.

Theorem 12.1.2. *Suppose that the characteristic function $f(t)$ satisfies for some $\lambda \geq 0$ the inequality*

(12.1.3) $$\liminf_{t \to \infty} \frac{-\log|f(t)|}{t^{1+\lambda}} > 0.$$

Then the corresponding distribution function $F(x)$ is an analytic distribution function. The domain of regularity of F contains the strip

$$S = \left\{ z = t + iy : |y| < \liminf_{t \to \infty} \frac{-\log|f(t)|}{t} \right\}.$$

ANALYTIC DISTRIBUTION FUNCTIONS

In this strip, F admits the representation

$$F(z) = F(0) + \frac{1}{2\pi i} \int_{-\infty}^{\infty} \frac{1 - e^{-itz}}{t} f(t) \, dt.$$

It is remarkable that the condition (12.1.3) is not necessary for the analyticity of F. It is possible to construct analytic distribution functions for which the limit (12.1.3) is equal to zero.

Interest in analytic distribution functions was revived recently. A brief discussion is also contained in T. Kawata (1972). B. Jesiak (1979) obtained the following result.

Theorem 12.1.3. Let $F(z)$ be an entire distribution function of order $\rho \leq 2$ or of order 2 and type 0. Then

$$\limsup_{y \to \infty} \frac{\log F(iy)}{y} > 0.$$

This result implies that an analytic distribution function cannot have order 1 and type 0.

Jesiak studied analytic distribution functions by means of the tail behaviour of the corresponding characteristic functions. He gave criteria which ensure that a distribution function $F(x)$ is an entire function and he studied the order and type of such functions. He also discussed the analytic properties of infinitely divisible distribution functions.

Theorem 12.1.4. Suppose that (12.1.3) holds for some $\lambda > 0$. Then $F(x)$ is an entire distribution function of order $\rho \leq 1 + \lambda^{-1}$. In the case where $\rho = 1 + \lambda^{-1}$ the type of $F(x)$ is finite.

These results contain the following special cases.

Theorem 12.1.5. Suppose that for some $\lambda > 0$ the relation

$$\liminf_{t \to \infty} \frac{\log[-\log|f(t)|]}{\log t} = 1 + \lambda$$

holds. Then $F(x)$ is an entire function of order $\rho \leq 1 + \lambda^{-1}$.

Theorem 12.1.6. A distribution function $F(x)$ is an entire distribution function if

$$\liminf_{t \to \infty} \frac{-\log|f(t)|}{t} = \infty.$$

The function $F(t)$ can be of finite or infinite order.

Jesiak also studied a special class of distribution functions for which the results formulated above can be sharpened. We introduce the following notations.

A distribution function $F(x)$ is said to belong to the class S^+ if its characteristic function $f(t)$ satisfies the following conditions:

(i) There exists a real x_0 such that $e^{-itx_0}f(t) \geq 0$ for all real t;

(ii) $\int_{-\infty}^{\infty} |f(t)|\, dt < \infty$.

It is easily seen that a distribution function $F \in S^+$ is symmetric with respect to x_0 and is absolutely continuous.

We write
$$W(t) = \int_t^\infty |f(u)|\, du$$
and
$$W_1(t) = \frac{\log[-\log W(t)]}{\log t}, \quad W_2^{(\lambda)}(t) = \frac{-\log W(t)}{t^{1+\lambda}} \quad (\lambda \geq 0).$$

Theorem 12.1.7. *Suppose that $F \in S^+$. Then:*

(1) *F is of order 1 and intermediate type if and only if $W(t) = 0$ for some $t > 0$.*

(2) *F is of order 1 and maximal type if and only if $W(t) > 0$ for all $t > 0$ and*
$$\lim_{t \to \infty} W_1(t) = \infty.$$

(3) *F is of order $\rho = 1 + \lambda^{-1}$ ($0 \leq \lambda < \infty$) and of type τ ($0 \leq \tau \leq \infty$, $\lambda > 0$) if and only if*
$$W(t) > 0 \text{ for all } t > 0, \quad \liminf_{t \to \infty} W_1(t) = 1 + \lambda;$$
$$\liminf_{t \to \infty} W_2^{(\lambda)}(t) = \begin{cases} c_\lambda \tau^{-\lambda} & \text{for } 0 < \lambda < \infty \\ \infty & \text{for } \lambda = 0, \end{cases}$$
where $c_\lambda = \lambda^\lambda/(1+\lambda)^{1+\lambda}$.

We now introduce a concept which is useful in the theory of analytic distribution functions [see Jesiak (1979)].

Let F be an analytic distribution function. A horizontal strip $S(s) = \{z: |\operatorname{Im} z| < s\}$, which belongs to the domain of regularity of F, is said to be a strip of boundedness if for any s_1 ($0 < s_1 < s$) there exists a constant $C(s_1)$, which may depend on s_1, such that $|F(z)| < C(s_1)$ for $z \in \bar{S}(s_1)$.

Theorem 12.1.8. *A distribution function $F(x)$ is analytic with a strip of boundedness if and only if*

(12.1.4) $\quad s_0 = \liminf_{t \to \infty} \dfrac{-\log|f(t)|}{t} > 0.$

Then $S(s_0)$ is a strip of boundedness for $F(x)$.

The existence of a strip of boundedness is also important for the study of convolutions.

Theorem 12.1.9. *Let $F_1(x)$ be an analytic distribution function with a strip of boundedness $S(s)$, and let F_2 be an arbitrary distribution function. Then the convolution $F = F_1 * F_2$ is an analytic distribution function with strip of boundedness $S(s)$, and $F(z)$ admits the representation by a convolution integral, namely*

$$F(z) = \int_{-\infty}^{\infty} F_1(z-u) \, dF_2(u) \qquad \text{for } z \in S(s).$$

Jesiak has also studied analytic distribution functions which are infinitely divisible. He extended earlier results about the class L obtained by Zolotarev (1963). In order to apply the previous statements we introduce the notation

$$N_\gamma(v) = N(\gamma v) - N(v), \qquad M_\gamma(v) = M(-v) - M(-\gamma v)$$

for $v > 0$, where N and M are the spectral functions of an infinitely divisible distribution.

Theorem 12.1.10. *Suppose that for some $\gamma > 1$ and $\lambda \geq 0$*

$$\liminf_{v \to +0} v^{1+\lambda} [N_\gamma(v) + M_\gamma(v)] > 0;$$

then $s_0 > 0$. Here s_0 is given by (12.1.4). This means that $F(x)$ is an analytic distribution function. In the case where $\lambda > 0$ the assertions of Theorem 12.1.4 remain true.

12.2 Continuation of distribution functions

This problem has been extensively treated in Rossberg, Jesiak & Siegel (1981) and was, according to Ibragimov (1977), first raised by Kolmogorov who conjectured that the normal distribution is an infinitely divisible distribution which is determined by its values on the negative half-axis. A proof of this uniqueness property of the normal distribution was given by Rossberg who obtained the following result.

Theorem 12.2.1. *Let $F(x)$ be an infinitely divisible distribution function and let $\Phi(x)$ be the normal distribution with mean 0 and variance 1. If $F(x) = \Phi(x)$ for $x < 0$ then $F(x) = \Phi(x)$ for all x.*

A generalization was given by M. Riedel (1975); see also Rossberg, Jesiak & Siegel (1981).

Theorem 12.2.2. *Let $F(x)$ be an infinitely divisible distribution which satisfies the relation*

$$\lim_{x \to -\infty} \frac{F(x)}{\Phi(x)} = 1.$$

Then $F(x) \equiv \Phi(x)$.

Theorem 12.2.3. Let $F(x)$ be an infinitely divisible distribution and suppose that the characteristic function $f(t)$ of $F(x)$ can be continued analytically into the upper [lower] half-plane. If the infinitely divisible distribution $G(x)$ coincides with $F(x)$ on the half-line $(-\infty, a)$ [on the half-line (a, ∞)] then it either becomes zero [or one] on some half-line or $F(x) = G(x)$ for all x.

This generalization of Theorem 12.2.1 is due to Ibragimov (1977).

Other classes of infinitely divisible distribution functions are also of interest; this subject is discussed in Rossberg, Jesiak & Siegel (1981). As an example we quote the following result concerning stable distribution functions. It is due to Rossberg & Jesiak (1978), Zolotarev (1978) and, in the final form cited here, to Jesiak (1979).

Theorem 12.2.4. Let F be a stable distribution function and let $\{x_j\}_{j=1}^{\infty}$ be a countable point set. Suppose that $F(x_i)$ ($i = 1, 2, \ldots$) is known and that the $F(x_i)$ are different from 0 or 1. Then $F(x)$ is uniquely defined.

An interesting result on the continuation of analytic distribution functions is due to Siegel.

Theorem 12.2.5. Let $F(x)$ be a distribution function with real characteristic function $f(t)$. Suppose that for a constant $B > 0$,

(1) $F(x) - F(-x) \leqslant Bx, \quad 0 \leqslant x < x_1;$
(2) $f(t) \geqslant 0 \quad$ for all real t.

Let S be a point set with limit point 0 and let G be a symmetric analytic distribution function such that

$$F(x) = G(x) \quad \text{for } x \in S.$$

Then $F \equiv G$.

As to further topics in this area we refer to Rossberg, Jesiak & Siegel (1981). These authors mention in their paper several problems which are not yet solved.

12.3 Limit theorems and restricted convergence

Let S be a subset of R_1 and let $\{F_n(x)\}$ be a sequence of distribution functions. Suppose that the sequence F_n converges weakly to a distribution function F on S. We write

(12.3.1) $\quad \lim_{n \to \infty} F_n(x) = F(x) \quad$ for $x \in S$,

and we then discuss restricted convergence of the sequence $F_n(x)$ to $F(x)$.

The question arises of how to describe situations in which restricted convergence implies weak convergence on R_1. These investigations led to a new type of limit theorems for sums of independent random variables.

ANALYTIC DISTRIBUTION FUNCTIONS

We mention a limit theorem of this type which is due to Rossberg & Siegel (1975).

Theorem 12.3.1. Let $\{X_v\}$ be a sequence of independently and identically distributed random variables; we write

(12.3.2) $\quad S_n = B_n^{-1}(X_1 + X_2 + \ldots + X_n) - A_n \quad B_n > 0$

and denote the distribution function of S_n by F_n. Suppose that for $x < 0$ the sequence $F_n(x)$ converges weakly to the standardized normal distribution $\Phi(x)$. Then the sequence $F_n(x)$ converges weakly to $\Phi(x)$ for all x.

In the following we put

$$\underline{F}(x) = \liminf_{n \to \infty} F_n(x), \quad \bar{F}(x) = \limsup_{n \to \infty} F_n(x).$$

The next theorem (due to Riedel (1977)) indicates that the asymptotic behaviour of $\underline{F}$ and $\bar{F}$ implies convergence and also the uniqueness of the limit distribution.

Theorem 12.3.2. Let $\{X_j\}_{j=1}^{\infty}$ be a sequence of independent and identically distributed random variables with common distribution $G(x)$. Suppose that

$$\lim_{x \to -\infty} \frac{\underline{F}(x)}{\Phi(r)} = \lim_{x \to \infty} \frac{\bar{F}(x)}{\Phi(x)} = 1.$$

Then

$$\operatorname*{Lim}_{n \to \infty} F_n(x) = \Phi(x).$$

We next give a condition which assures that a limit distribution is stable. Let $X_1, X_2, \ldots, X_n$ be independently and identically distributed random variables with common distribution function $G(x)$, and let S_n be defined by (12.3.2) and denote again the distribution function of S_n by $F_n(x)$. Rossberg (1979) studied this case in connection with the concept of restricted convergence, and he obtained the following result.

Theorem 12.3.3. Suppose that for a certain distribution function $G(x)$ one has

$$\operatorname*{Lim}_{n \to \infty} F_n(x) = F(x) \quad \text{for } x \leq 0$$

and assume that $F(x)$ is a monotone function such that $F(-\infty) = 0$, $F(x) > 0$ for $x \leq 0$. Then

$$\operatorname*{Lim}_{n \to \infty} F_n(x) = F_\alpha(x),$$

where $F_\alpha(x)$ is a stable distribution with exponent α $(0 < \alpha \leq 2)$.

The proofs of such theorems split into two parts. In the first, we need to show that every subsequence $\{F_n'\}$ converging weakly on the whole line has a non-defective limit distribution which is (this is the second step) uniquely defined by the values given on the set S. The second problem is obviously a continuation problem and the results sketched in Section 12.2 can be applied. As to the first, we need criteria of relative compactness. We say that $\{F_n(x)\}$ is relatively compact if every subsequence $\{F_n'\}$ contains a subsequence $\{F_n''\}$ which converges completely to a proper distribution function.

We have the following useful criterion which was also needed in the proofs of Theorems 12.3.1, 12.3.2 and 12.3.3.

Theorem 12.3.4. *Let $\{F_n\}$ be a sequence of distribution functions of independent and identically distributed random variables, and suppose that $\underline{F}(x) > 0$ for all x, $\bar{F}(-\infty) = 0$. Then the sequence $\{F_n\}$ is relatively compact.*

It is also possible to obtain limit theorems of this kind for triangular arrays. Let $\{X_{nk}\}$ $[1 \leqslant k \leqslant k_n,\ \lim_{n \to \infty} k_n = \infty,\ n = 1, 2, \ldots]$ be a triangular array of random variables and denote the distribution function of X_{nk} by F_{nk}. Let

$$T_n = \sum_{j=1}^{k_n} X_{nj} - A_n$$

and write $\underline{G}_n(x)$ for the distribution function of the random variable T_n.

Here $\{A_n\}$ is a sequence of constants.

Theorem 12.3.5. *The sequence $\{G_n\}$ is relatively compact if and only if*

(i) $\underline{G}(x_0) > 0$ *for some real* x_0, $\bar{G}(-\infty) = 0$, *and*

(ii) $\sup_n \sum_{k=1}^{k_n} P(x_{nk} - m_{nk} \geqslant x) = o(1) \qquad (x \to \infty)$.

Here m_{nk} is a median of X_{nk}.

This and other related results can be found in Rossberg, Jesiak & Siegel (1981).

13 Metrics in the space of distribution functions

We saw in Chapter 3, Section 1, that the set of all distribution functions can be made into a metric space by defining a distance between two distribution functions. We also introduced two very important metrics, namely the uniform (Kolmogorov) metric and the Lévy metric, and we obtained some results concerning their connection.

These two metrics are essential tools in the study of probabilistic stability theorems;(*) however, they are not the only metrics which one can define on the space of distribution functions. Metrics in this space are important in connection with a variety of studies. As examples we mention investigations of convergence, of approximations, of characterizations of distributions, and studies of stability problems. However, the discussion of these applications exceeds the scope of this book; they would in fact provide enough material for a separate monograph. The aim of this chapter is more modest, and we shall give here only a brief outline of the theory of metrics in the space of distribution functions.

13.1 Properties of metrics defined on a set X; comparison of metrics

In order to compare different metrics we introduce the following definitions.

Let X be a space on which two metrics $\mu_1(x, y)$ and $\mu_2(x, y)$ are defined $(x, y \in X)$. We say that the metric μ_2 is stronger than the metric μ_1 (alternatively μ_1 is weaker than μ_2) if any sequence $\{x_n\}$ of points of X which converges to a point x in the metric μ_2 also converges(†) in the metric μ_1. We then write $\mu_1 \leqslant \mu_2$. If $\mu_1 \leqslant \mu_2$ and if there exists a sequence $\{y_n\}$ and an element y such that $\mu_1(y_n, y) \to 0$ but $\lim_{n \to \infty} \mu_2(y_n, y) \neq 0$ then we say that μ_1 is strictly weaker than μ_2 [μ_2 is strictly stronger than μ_1] and write $\mu_1 < \mu_2$ [or $\mu_2 > \mu_1$].

If $\mu_1 \leqslant \mu_2$ and $\mu_2 \leqslant \mu_1$ we say that μ_1 and μ_2 are equivalent and write $\mu_1 \sim \mu_2$. If the metrics μ_1 and μ_2 are not equivalent then we write $\mu_1 \not\sim \mu_2$.

Remark. There exist metrics which are not comparable with each other. We consider the following example:

(*) A stability theorem is always associated with a theorem in probability theory. One supposes that the assumptions of the theorem are not exactly but only approximately satisfied, and one examines how this affects the conclusions.

(†) We say that x_n converges to x_0 in a metric μ if $\lim_{n \to \infty} \mu(x_n, x_0) = 0$ and we write $\mu(x_n, x_0) \to 0$.

Let $X = (0, \infty)$ be the set of positive real numbers and define the two metrics $\mu_1(x, y) = |x - y|$, $\mu_2(x, y) = \left| \frac{1}{x} - \frac{1}{y} \right|$ on X. One can easily show that $\mu_1 \sim \mu_2$ but that there exist sequences $\{x_n\}$ and $\{y_n\}$ in X such that $\mu_1(x_n, y_n) \geq \epsilon > 0$, while $\mu_2(x_n, y_n) \to 0$. In such a case it is not possible to estimate, uniformly on X, one metric by the other.

Theorem 13.1.1. Let μ_1 and μ_2 be two equivalent metrics defined on X. Let $A \subset X$ and suppose that A is compact(*) in X in these metrics. Then there exist functions $\phi_1(\epsilon)$ and $\phi_2(\epsilon)$, depending on A, such that

$$\lim_{\epsilon \to 0} \phi_1(\epsilon) = 0, \quad \lim_{\epsilon \to 0} \phi_2(\epsilon) = 0$$

which satisfy the inequalities

(13.1.1) $\begin{cases} \mu_1(x,y) \leq \phi_1(\mu_2(x,y)) \\ \mu_2(x,y) \leq \phi_2(\mu_1(x,y)). \end{cases}$

To prove this we construct the function

$$\phi(\epsilon) = \sup\{\mu_1(x, y) \quad \text{such that } x \in A, y \in X, \mu_2(x, y) \leq \epsilon\}.$$

We assume tentatively that

$$\lim_{\epsilon \to 0} \phi(\epsilon) \neq 0.$$

Then there exists a $\delta > 0$ and sequences $\{\epsilon_n\}, \{x_n\}, \{y_n\}$ such that $\lim_{n \to \infty} \epsilon_n = 0$, $\{x_n\} \subset A, \{y_n\} \subset X$ and

(13.1.2a) $\mu_2(x_n, y_n) \leq \epsilon_n$

while

(13.1.2b) $\mu_1(x_n, y_n) \geq \delta > 0$.

In view of the compactness of A there exists a subsequence $\{x_{n_j}\}$ which converges in both metrics μ_1 and μ_2 to some $x \in A$. On account of the triangle inequality one has

$$\mu_2(y_{n_j}, x) \leq \mu_2(y_{n_j}, x_{n_j}) + \mu_2(x_{n_j}, x) \to 0.$$

Therefore

$$\mu_2(y_{n_j}, x) \to 0, \quad \text{so that also } \mu_1(y_{n_j}, x) \to 0.$$

This contradicts (13.1.2b) so that $\lim_{\epsilon \to 0} \phi(\epsilon) = 0$. The second inequality is treated in the same way.

(*) $A \subset X$ is said to be compact in X in the metric μ if it is possible to select from every sequence $\{x_n\} \subset A$ a subsequence which converges in the metric μ to some $x \in A$.

METRICS IN THE SPACE OF DISTRIBUTION FUNCTIONS 133

In the following we assume that all metrics μ considered are defined on the space X and satisfy the inequality $\mu \leq 1$. This can be done without loss of generality, since one could use instead of μ the equivalent metric $\nu = \mu/(1 + \mu)$. We also introduce the metric

$$d_0(x, y) = \begin{cases} 0 & \text{if } x = y \\ 1 & \text{if } x \neq y. \end{cases}$$

Clearly, every metric defined on X is weaker than d_0.

Let μ be a metric such that $\mu \leq 1$ and let $A \subset X$. Then we define the metric

$$\widetilde{\mu}(x, y | A, X) = \widetilde{\mu}(x, y) = \begin{cases} \mu(x, y) & \text{if } x, y \in A \\ d_0(x, y) & \text{otherwise}. \end{cases}$$

Theorem 13.1.2. Let μ be a metric such that $\mu \not\sim 0$. Then there exists a metric ν such that $\mu \prec \nu$ and $\nu \not\sim d_0$.

Since μ is not equivalent to d_0 there exists a sequence $\{x_n\}$ consisting of different elements and an $x_0 \in X$ such that $\mu(x_n, x_0) \to 0$. Let A be the sequence $\{x_{2n}\}$ and put $\nu(x, y) = \widetilde{\mu}(x, y | A, x)$. It is clear that $\nu \not\sim d_0$ and that only the subsequence $\{x_{2n}\}$ converges in the metric ν. But $\nu(x_{2n+1}, x_0) = 1$, so that $\mu \prec \nu$ and the theorem is proved.

Lemma 13.1.1. Let $\mu_1 \prec \mu_2$ and let the sequence $\{x_n\}$ be such that $\mu_1(x_n, x_0) \to 0$ while $\mu_2(x_n, x_0) \geq \epsilon > 0$. Then there exists a continuous function(*) $f(x)$ on X such that $f(x_{2n}) = 1, f(x_{2n+1}) = 0$ for all non-negative integers n.

Proof. Let $k \geq 0$ be an integer and let $S_{2k} = \{x : \mu_2(x_{2k}, x) \leq \epsilon_k\}$. For any non-negative integer k there exists an $\epsilon_k > 0$ such that $x_j \notin S_{2k}$ provided $j \neq 2k$. We define

$$f(x) = \begin{cases} 1 - \dfrac{\mu_2(x_{2k}, x)}{\epsilon_k} & \text{if } x \in S_{2k} \\ 0 & \text{if } x \notin \bigcup_{k=0}^{\infty} S_{2k}. \end{cases}$$

Then $f(x_{2k}) = 1$ and $f(x_{2k+1}) = 0$ since $x_{2k+1} \notin S_{2l}$ for all l. Moreover $f(x)$ is continuous in the metric μ_2.

The next theorem generalizes Theorem 13.1.2.

Theorem 13.1.3. Let $\mu_1 \prec \mu_2$; then there exists a metric ν such that $\mu_1 \prec \nu \prec \mu_2$.

According to our assumption there exists a sequence $\{x_n\}$ such that $\mu_1(x_n, x_0) \to 0$ while $\mu_2(x_n, x_0) \geq \epsilon > 0$. Let

$$\nu(x, y) = \mu_1(x, y) + |f(x) - f(y)|,$$

(*) Continuity is here defined in the metric μ_2.

where $f(x)$ is a function whose existence is assured and whose properties are described in Lemma 13.1.1. Then v is a metric and $\mu_1 \leqslant v \leqslant \mu_2$. Since $v(x_{2n}, x_0) = \mu_1(x_{2n}, x_0) \to 0$ while $\mu_2(x_{2n}, x_0) \geqslant \epsilon > 0$, we see that $v(x_{2n+1}, x_0) \geqslant 1$ but $\mu(x_{2n+1}, x_0) \to 0$. Therefore $\mu_1 \prec v \prec \mu_2$.

Theorem 13.1.4. *Let μ be a metric given on X and suppose that X is not compact in μ. Then there exists a metric v, defined on X, such that $v \prec \mu$.*

For the proof we refer to Senatov (1977). His paper is also the source of the preceding statements.

Let

$$\phi_1(x) = \begin{cases} -x & \text{if } x < 0 \\ \dfrac{x}{x+1} & \text{if } x \geqslant 0 \end{cases}$$

$$\phi_2(x) = \begin{cases} -x & \text{if } x < 0 \\ \dfrac{x}{x+2} & \text{if } x \geqslant 0 \end{cases}$$

be two functions of the real variable x. We write

(13.1.3) $\quad d(x, y) = |x - y|$

(13.1.4) $\quad \eta(x, y) = |\phi_1(x) - \phi_1(y)| + |\phi_2(x) - \phi_2(y)|.$

It is easily seen that these functions are metrics on the set R of real numbers and that $\eta \leqslant d$. But $\eta \prec d$ since $\eta(-1, n) \to 0$.

13.2 The case where X is the set of distribution functions

V. V. Senatov (1977) used the metric (13.1.4) to construct a metric on the set of all distribution functions which is weaker than the Lévy metric.

Let $\mathscr{F}$ be the set of all non-decreasing right continuous functions $F(x)$ such that $0 \leqslant F(x) \leqslant 1$ and define for $T > 0$

$$F_T(x) = \begin{cases} 0 & \text{if } x < -T \\ F(x) & \text{if } -T \leqslant x < T \\ 1 & \text{if } x \geqslant T. \end{cases}$$

Theorem 13.2.1. *Let $F(x)$ and $G(x)$ be two arbitrary distribution functions and suppose that the metric μ is defined on the space of all distribution functions and has the following property: $\mu(F_T, G_T)$ is non-decreasing in T. Then there does not exist a metric on the space of distribution functions which has this property and which is weaker than the Lévy metric.*

For the proof we refer to Senatov (1977).

METRICS IN THE SPACE OF DISTRIBUTION FUNCTIONS 135

We say that a function $H(x)$ lies between the functions $F(x)$ and $G(x)$ if either $F(x) \leq H(x) \leq G(x)$ or $G(x) \leq H(x) \leq F(x)$ for any value x of the argument.

Let $F(x)$, $G(x)$ and $H(x)$ be distribution functions and suppose that H lies between F and G. A metric μ defined on the space of distribution functions is said to be monotone if $\mu(F, H) \leq \mu(F, G)$ for any distribution function such that H lies between F and G.

Theorem 13.2.2. *The weakest metric in the class of monotone metrics is the Lévy metric.*

Theorem 13.2.3. *Let $\mathcal{M}$ be the class of monotone metrics defined on the set of distribution functions which satisfy the following conditions:*

(i) $\mu(F(x), G(x)) = \mu(F(ax + b), G(ax + b))$ *for any $\mu \in \mathcal{M}$, any distribution functions F and G and $a > 0$, b real;*

(ii) $\mu(F_T, G_T)$ *is non-decreasing in T.*

Then $\mathcal{M}$ contains a weakest metric which is (up to equivalence) the uniform (Kolmogorov) metric.

The last two theorems are due to Senatov (1977). In his paper he gave a partial proof of Theorem 13.2.2 and stated Theorem 13.2.3 without proof.

Zolotarev (1977) studied continuity problems in probability theory and considered metrics — which he called ideal metrics — on the space of probability measures on a Banach space. As an example we give here the definition of an ideal metric of order s defined on the one-dimensional space R_1.

Let F and G be two distribution functions and let $s > 0$, $K > 0$ and write D_s^K for the class of complex-valued functions which have $m = [s]$ derivatives such that

$$|\phi^{(m)}(x) - \phi^{(m)}(y)| \leq K |x - y|^\alpha.$$

Here $\alpha = s - m$, $[s]$ is the integer part of m. The ideal metric ζ_s^K is then given by

$$\zeta_s^K(F, G) = \sup_{\phi \in D_s^K} \left| \int_{-\infty}^\infty \phi(x) \, d \left[F(x) - G(x) \right] \right|.$$

13.3 Metrics on the set of characteristic functions

The set of characteristic functions can be made into a metric space. In this section we define a distance between characteristic functions and discuss its connection with the Lévy metric.

Let F and G be two distribution functions with characteristic functions f and g respectively. Zolotarev & Senatov (1975) defined the distance between two characteristic functions f and g by the formula

$$\lambda(f, g) = \min_{T > 0} \max_{|t| \leq T} \left[|f(t) - g(t)|, \frac{1}{T} \right].$$

They showed that the metric λ is equivalent to the Lévy metric in the following sense.

Let $\{F_n\}$ be a sequence of distribution functions, let F be a distribution, and denote the corresponding characteristic functions by $\{f_n\}$ and f.

The fact that the sequence $\{F_n\}$ converges weakly to F (i.e. $\lim_{n \to \infty} L(F_n, F) = 0$) implies that $\lim_{n \to \infty} \lambda(f_n, f) = 0$ and vice versa.

These authors also gave estimates of λ in terms of L and of L in terms of λ.

In order to present these estimates it is convenient to introduce the following notations.

Let $\xi > 0$ and write $T_F(\xi)$ and $T_G(\xi)$ for the tails of the distribution functions from the point ξ onward, i.e. $T_F(\xi) = 1 - F(\xi) + F(-\xi)$; $T_G(\xi)$ is defined in the same way. Let $u(\xi) = \min[T_F(\xi), T_G(\xi)]$. We can now formulate the estimates.

Theorem 13.3.1. *The distance λ may be estimated from below, using the Lévy distance, by*

$$2[\lambda(f, g)]^2 \geq [4\xi + 4L(F, G) + 1] L(F, G),$$

where ξ is a positive number such that

$$4L(F, G) + 2u(\xi) \leq [\xi + L(F, G)] \left[\frac{2L(F, G)}{4\xi + 4L(F, G) + 1}\right]^{1/2}.$$

Theorem 13.3.2. *The Lévy distance $L(F, G)$ may be estimated using the distance $\lambda(f, g)$ by*

$$L(F, G) \leq 8\lambda(f, g) \log[Y/\lambda(f, g)].$$

Here Y is a number satisfying the condition

$$Y \geq e, u(Y/e) \leq \lambda(f, g) \log[Y/\lambda(f, g)].$$

For the proof the reader is referred to the paper by Zolotarev & Senatov (1975).

Let $F_1(x)$ and $F_2(x)$ be two distribution functions with characteristic functions $f_1(t)$ and $f_2(t)$ respectively. F. J. Dyson (1953) constructed an example to show that it is not possible to find for any arbitrary $\delta > 0$ an $\epsilon > 0$ which depends only on δ but is independent of x and $F_1(x)$ [$F_2(x)$] such that

(13.3.1) $|f_1(t) - f_2(t)| < \epsilon$ for all x

implies

(13.3.2) $|F_1(x) - F_2(x)| < \delta$.

We give Dyson's example:

METRICS IN THE SPACE OF DISTRIBUTION FUNCTIONS

Let $b > a > 0$ and let

$$F_1(x) = \begin{cases} \dfrac{1}{2}\log\left(\dfrac{x^2 + b^2}{x^2 + a^2}\right) \Big/ \log\dfrac{b}{a} & \text{if } x \leq 0 \\ 1 & \text{if } x \geq 0 \end{cases}$$

and

$$F_2(x) = 1 - F_1(-x).$$

Then, for all x.

(13.3.3) $\quad F_1(x) - F_2(x) = \dfrac{1}{2}\log\left(\dfrac{x^2 + b^2}{x^2 + a^2}\right) \Big/ \log\dfrac{b}{a}.$

Hence

$$F_1(0) - F_2(0) = 1.$$

It follows from (13.3.3) that

$$f_1(t) - f_2(t) = i\pi \, \dfrac{t}{|t|} \, [e^{-a|t|} - e^{-b|t|}] / \log\dfrac{b}{a}.$$

Therefore

(13.3.4) $\quad |f_1(t) - f_2(t)| < \pi / \log\dfrac{b}{a}.$

Since the right-hand side of (13.3.4) can be made arbitrarily small by taking b sufficiently large, (13.3.1) can be satisfied for any $\epsilon > 0$, but (13.3.2) is false for $\delta = 1$.

A second example, due to D. Szász but published in the paper of Zolotarev & Senatov (1975), shows that there exist distribution functions F and G, with characteristic functions $f(t)$ and $g(t)$ respectively, such that even for $L(F, G)$ small, $\lambda(f, g)$ is not small.

Let $F(x) = \dfrac{1}{n} \sum_{j=0}^{n-1} \epsilon(x - 2y)$ and $G(x) = F(x - n)$. Then

(13.3.5) $\quad \begin{cases} f(t) = \dfrac{1}{n} \, \dfrac{1 - \exp(2itn^2)}{1 - \exp(2itn)} \\ g(t) = f(t) \, e^{itn}, \end{cases}$

so that

$$w(t) = f(t) - g(t) = f(t)[1 - e^{itn}].$$

Hence we see that

$$(13.3.6) \quad w(t) = f(t) - g(t) = \frac{1}{n} \frac{1 - \exp(2itn^2)}{1 + \exp(itn)}.$$

By means of elementary computations we see that

$$(13.3.7a) \quad \lim_{\tau \to \pi/n} [f(t) - g(t)] = 2,$$

$$(13.3.7b) \quad \max_{|\tau| \leq 1} |w(t)| = 2,$$

so that

$$(13.3.8) \quad \lambda(f, g) = 1.$$

It follows from the definition of $F(x)$ and $G(x)$ that

$$(13.3.9) \quad L(F, G) = \frac{1}{n}.$$

Therefore $L(F, G)$ can be made arbitrarily small by a suitable choice of n, while $\lambda(f, g) = 1$.

14 Ridge functions

We saw in Lukacs (1970) (Theorem 7.1.2) that analytic characteristic functions have the ridge property: this means that every analytic characteristic function satisfies (in its strip of regularity) the inequality $|f(t+iy)| \leqslant f(iy)$. In this chapter we study ridge functions and show that this class of functions contains the set of characteristic functions as a proper subclass. This investigation will not only yield information about an interesting class of analytic functions but will also indicate which results on analytic characteristic functions depend on the ridge property.

14.1 Definition of ridge functions and the existence of ridge functions which are not analytic characteristic functions

We consider a complex-valued function $f(z)$ of the complex variable $z = t + iy$ which is regular in the horizontal strip

(14.1.1) $\quad \alpha < y < \beta, \quad |t| < \infty.$

This function is said to be a ridge function[*] if

(14.1.2) $\quad |f(t+iy)| \leqslant |f(iy)|.$

In the cases $\alpha = -\infty$ and $\beta = +\infty$ we speak of an entire ridge function. The strip (14.1.1) is called the strip of regularity of the ridge function $f(z)$.

Remark. We do not assume that the strip (14.1.1) contains the real axis in its interior. However, we shall mostly study ridge functions for which the point $z = 0$ is either in the interior or on the boundary of the strip (14.1.1).

Theorem 14.1.1. Products of ridge functions are ridge functions. Arbitrary positive powers of ridge functions are ridge functions.

A ridge function $f(z)$ is said to be standardized if the point $z = 0$ is either in the interior or on the boundary of its strip of regularity and if $f(0) = 1$.

Theorem 14.1.2. All analytic characteristic functions and all boundary characteristic functions are ridge functions. The converse is not true, there exist ridge functions which are not characteristic functions.

---

[*] Functions of this class were first introduced by D. Dugué (1951a) though without using the term "ridge function".

The first part of the statement is trivial. We prove the second part of the theorem by constructing a suitable example.

Example. Let $a > 1$ and

(14.1.3) $\quad f(z) = \exp\left[-\dfrac{1}{a + e^{iz}} + \dfrac{1}{a+1}\right].$

By an elementary computation we see that

$$|f(t + iy)| = \exp\left\{\dfrac{1}{a+1} - \dfrac{a + e^{-y}\cos t}{a^2 + e^{-2y} + 2a\, e^{-y}\cos t}\right\}.$$

We note that $(a + e^{-y}x)[a^2 + e^{-2y} + 2a\, e^{-y}x]^{-1}$ is a decreasing function of x, provided that $y > -\log a$. We then have

$$|f(t + iy)| \leq \exp\left(\dfrac{1}{a+1} - \dfrac{1}{a + e^{-y}}\right) = f(iy).$$

Therefore $f(z) = f(t + iy)$ is a standardized ridge function. This function has singularities at the points

$$z_k = (2k + 1)\pi - i\log a \qquad (k = 0, \pm 1, \pm 2, \ldots)$$

and it is regular in the region $\{|\text{Re}\, z| < \pi\}$. This region contains the whole imaginary axis. If $f(z)$ were an analytic characteristic function then its strip of regularity would be the whole plane. This is impossible on account of the singularities z_k. Thus, there are standardized ridge functions which are not characteristic functions. This example is due to Dugué (1951a) and is also presented in Linnik & Ostrovskii (1977).

The next example shows that there are entire ridge functions which are not characteristic functions.

We consider again the polynomial

(14.1.4) $\quad P(x) = 1 + a_1 x - a_2 x^2 + a_3 x^3 + a_4 x^4,$

where all $a_j > 0$. We select the coefficients in (14.1.4) in such a way that $[P(x)]^2$ as well as $[P(x)]^3$ have no negative coefficients.[*] It follows that all coefficients of $\exp[P(x)]$ are non-negative if and only if

(14.1.5) $\quad a_2 \leq a_1^2/2.$

Therefore

(14.1.6) $\quad g(t) = \exp[P(e^{it}) - P(1)]$

is a characteristic function if (14.1.5) is satisfied, and it is not a characteristic

---

[*] This can be accomplished if $a_1 = a_3 = a_4 = 1$, $a_2 = \frac{1}{4}$.

RIDGE FUNCTIONS

function if

(14.1.7) $a_2 > a_1^2/2$.

We note that $g(t)$ is an entire function of infinite order.
We now introduce the function

(14.1.8) $g_\alpha(z) = [g(z)]^\alpha = \exp[\alpha P(e^{it}) - \alpha P(1)]$, $(z = t + iy)$

or

$$g_\alpha(t) = e^{-\alpha P(1)} \exp[\alpha P(e^{it})].$$

In view of the preceding discussion we see that the expansion of $\exp[\alpha P(x)]$ has no negative coefficients if

(14.1.9a) $\alpha \geq 2a_2/a_1^2$,

but this expansion has a negative coefficient if

(14.1.9b) $\alpha < 2a_2/a_1^2$.

This means that $g_\alpha(t)$ is an entire characteristic function if (14.1.9a) holds, and it is therefore a ridge function.

We see from (14.1.8) that

$$g_1(z) = g(z), \quad (z = t + iy)$$

and we therefore examine the function $g(z)$.

According to (14.1.6) one has

$$e^{P(1)} g(z) = \exp[P(e^{iz})],$$

where

$$\operatorname{Re} P(e^{iz}) = \operatorname{Re} P[\exp(it - y)] \leq P(e^{-y}),$$

so that

$$|\exp[P(e^{iz})]| \leq \exp[P(e^{-y})].$$

Therefore

(14.1.10) $|e^{P(1)} g(z)| \leq \exp[P(e^{-y})] = \exp[P(1)] g(iy)$.

Hence $g(z)$ is a ridge function. It follows from (14.1.8) and Theorem 14.1.1 that $g_\alpha(z)$ is an entire ridge function for any positive α. However, it is not a characteristic function for $\alpha < 2a_2/a_1^2$.

It is also possible to construct standardized ridge functions which are entire functions of finite order but are not characteristic functions. An example is given in A. M. Kagan, Yu. V. Linnik & C. R. Rao (1973).

14.2 Elementary properties of ridge functions

Theorem 14.2.1. Let $z = t + iy$ and suppose that $f(z)$ is a ridge function in the strip $\alpha < y < \beta$. Then

(a) $f(iy) \neq 0$ for $\alpha < y < \beta$;
(b) $\arg f(iy) = $ const. for $\alpha < y < \beta$;
(c) $B(y) = \log|f(iy)|$ is convex in $\alpha < y < \beta$.

These properties correspond to properties of analytic characteristic functions. They are proved in the same way as the corresponding results on analytic characteristic functions.

Corollary 1 to Theorem 14.2.1. Let $f(z)$ be a standardized ridge function in $\alpha \leq y \leq \beta$ ($\alpha \leq 0 \leq \beta$), then $f(iy) > 0$ in this strip. This follows from the fact that $\arg f(iy) \neq 0$ and that $\arg f(iy) = \arg f(0) \equiv 0 \pmod{2\pi}$.

Corollary 2 to Theorem 14.2.1. Let $f(z)$ be an entire, standardized ridge function which has no zeros. Then $f(z) = \exp[\phi(z)]$ where $\phi(z)$ is an entire function which is real-valued on the imaginary axis and where $\phi(0) = 0$.

An entire function without zeros has the form $f(z) = \exp[\phi(z)]$. By Corollary 1, $f(z)$ is positive on the imaginary axis; hence $\text{Im} f(z) = 2\pi k(z)$, where $k(z)$ assumes only integer values. It follows from $\phi(0) = 0$ that $k(z) \equiv 0$.

Corollary 3 to Theorem 14.2.1. The zeros of a ridge function in its strip of regularity are located symmetrically about the y-axis.

This follows from Schwarz's reflection principle (see Appendix A2).

Corollary 4 to Theorem 14.2.1. Let $f(z)$ be an entire ridge function not equal to a constant. Then

$$M(r, f) = \max\{f(ir), f(-ir)\}.$$

Proof. Suppose that z_0 is a point on the circle $|z| = r$ such that $M(r, f) = |f(z_0)|$. Then $M(r, f) = |f(z_0)| \leq |f(i \, \text{Im} \, z_0)|$. If the point z_0 is different from $\pm ir$ then $i \, \text{Im} \, z_0$ is in the interior of $|z| < r$, which contradicts the maximum modulus principle.

Corollary 5 to Theorem 14.2.1. Let $f(z)$ be an entire standardized ridge function. Then

$$\lim_{y \to \infty} \frac{\log f(-iy)}{y} \quad \text{and} \quad \lim_{y \to -\infty} \frac{\log f(iy)}{y}$$

exist (possibly are infinite).

This follows from the fact that the function $\phi(y) = \log f(iy)$ is convex in the strip, and $\log f(0) = 0$ and is increasing for $y > 0$.

Theorem 14.2.2. Let $f(z)$ be an entire ridge function of finite order ρ; then $\rho \leq 2$.

Theorem 14.2.2 corresponds to Theorem 7.3.5 of Lukacs (1970) and is proved in a similar way.

14.3 Factorization of ridge functions

In the following we consider standardized ridge functions with the strip of regularity $-\alpha < \operatorname{Im} z < \beta$ ($\alpha > 0, \beta > 0$).

The ridge function $f_1(z)$ is said to be a ridge component (ridge factor) of the ridge function $f(z)$ if there exists a standardized ridge function $f_2(z)$ such that the equation

(14.3.1) $\quad f(z) = f_1(z) f_2(z)$

holds in the strip of regularity of $f(z)$. As in the case of characteristic functions, we consider only non-degenerate ridge components. Components which have the form e^{iaz} (a real) are called improper components.

A non-degenerate ridge function is said to be indecomposable if it has only improper ridge components.

Let $f(z)$ be a ridge function with the strip of regularity $-\alpha < \operatorname{Im} z < \beta$ ($\alpha > 0, \beta > 0$). In order that the function $[f(z)]^\lambda$ should be regular in the strip $-\alpha < \operatorname{Im} z < \beta$ for any $\lambda > 0$, it is necessary to require that $f(z)$ should have no zeros in its strip of regularity. If this condition is satisfied then one sees from Theorem 14.1.1 that $[f(z)]^\lambda$ is also a ridge function for any $\lambda > 0$. It is therefore justified to use the following definition:

A ridge function is said to be infinitely divisible if it has no zeros in its strip of regularity.

This definition differs from the corresponding definition for characteristic functions since an analytic characteristic function which has no zeros in its strip of regularity is not necessarily infinitely divisible.(*) This circumstance distinguishes the arithmetic of characteristic functions from the arithmetic of ridge functions.

In spite of this difference between ridge functions and characteristic functions, theorems corresponding to Theorems 6.2.1 and 6.2.2 of Lukacs (1970) are valid also for ridge functions.

Let $f(z)$ be a standardized ridge function with the strip of regularity $S = \{z : -\alpha \leq \operatorname{Im} z \leq \beta\}$ and let $R \leq \min(\alpha, \beta)$. Then the strip $S_R = \{z : |\operatorname{Im} z| < R\}$ is contained in S.

Theorem 14.3.1. A ridge function which has no indecomposable factor is infinitely divisible.

Theorem 14.3.2. Every ridge function can be represented as the product of at most two ridge functions such that one has no indecomposable factor while the other is the product of a finite or denumerable number of indecomposable ridge factors which converge absolutely and uniformly on any compact set of S_R.

(*) For example, the function (14.1.6), where (14.1.5) holds, is a characteristic function without zeros but is not infinitely divisible since (14.1.9a) does not hold for $\alpha = 1/n$ and large n.

For the derivation of these theorems one needs some properties concerning the decomposition of ridge functions. We formulate these as lemmas.

Lemma 14.3.1. *Let $f(z)$ be a standardized ridge function and let $k(z)$ be a ridge component (factor) of $f(z)$. Then $k(z)$ can be written in the form $k(z) = e^{-i\alpha z} k_0(z)$, where α is a real number and k_0 is a ridge factor of f such that $k_0'(0) = 0$. If k is indecomposable then k_0 is also indecomposable.*

If k is a ridge function one has $|k(t)| \leqslant k(0)$ for all real t, so that $k'(0) = 0$. Then $\alpha = i \operatorname{Re} k'(0)$ is real and the statement follows.

Lemma 14.3.2. *Let $f(z)$ be a standardized ridge function and consider the set $[g(z)]$ of all standardized ridge components of $f(z)$ such that $g'(0) = 0$. The functions of this family are uniformly bounded in the strip of regularity of $f(z)$.*

To each function $g(z)$ there exists a ridge function $k(z)$ such that $f(z) = g(z) k(z)$. We put

$$a = if'(0).$$

Since $g(0) = 1$ and $g'(0) = 0$ one sees that $f'(0) = k'(0)$. Let $k_1(z) = e^{iaz}k(z)$, then it follows that $k_1'(0) = 0$. Therefore $k_1(iy) \geqslant 1$ in the strip S_R. Hence

$$g(iy) \leqslant g(iy) k_1(iy) = e^{-ay} k(iy) g(iy) = e^{-ay} f(iy)$$

and

$$|g(z)| \leqslant e^{-ay} f(iy).$$

This is the statement of the lemma.

Lemma 14.3.3. *Let $f(z)$ be a standardized ridge function and $\{g_n(z)\}$ a sequence of its standardized ridge factors. Suppose that*

(14.3.2) $\quad \lim_{n \to \infty} g_n(z) = g(z)$

uniformly on every compact set of the strip S_R. Then $g(z)$ is a ridge component of $f(z)$.

Proof. It follows from a theorem of Weierstrass [see Markuševič (1965), vol. 1, p. 333] that $g(z)$ is regular in the strip S_R. Since the $g_n(z)$ are standardized ridge functions $g(z)$ is also a standardized ridge function. We consider the function

(14.3.3) $\quad k(z) = f(z)/g(z).$

This function is meromorphic in S_R; to prove the lemma we must show that $k(z)$ is a ridge function.

Since $g(z)$ is a ridge function it does not vanish on the segment of the imaginary axis which is located in the interior of S_R. Let

$$0 < \alpha_1 < \alpha; \quad 0 < \beta_1 < \beta \quad \text{and} \quad S_1 = \{z: -\alpha_1 < \operatorname{Im} z < \beta_1\}.$$

RIDGE FUNCTIONS   145

Let K be a compact set in S_1 such that it contains the segment $-\alpha_1 < \text{Im } z < \beta_1$ but no zero of the function $g(z)$. It follows from a theorem of Hurwitz [see Markushevič (1965), vol. 2, p. 49] that the $g_n(z)$ have no zeros in K, provided n is sufficiently large. We consider the sequence

$$k_n(z) = f(z)/g_n(z).$$

Since $g_n(z)$ are, by assumption, ridge components of $f(z)$, we see that $k_n(z)$ is a ridge function, so that

(14.3.4) $|k_n(t + iy)| \leq k_n(iy)$.

The sequence $[k_n(z)]$ converges uniformly to $k(z)$ on K, so that $k(z)$ is analytic on K. Since K is arbitrary, one sees that

(14.3.5) $|k(t + iy)| \leq k(iy)$

holds in every strip S_1. Since $k(z)$ has no poles on the segment of the imaginary axis located in the interior of S_1, it follows from (14.3.5) that $k(z)$ has no poles in S_1 and is a ridge function.

The proof of Theorems 14.3.1 and 14.3.2 is effected in a way which is similar to the proofs of Theorems 6.2.1 and 6.2.2 in Lukacs (1970).

We introduce the functional, applicable to ridge functions,

(14.3.6) $M_r(f) = \sup\limits_{|y| < r} \log f(iy),$

where $|r| < R$. This functional corresponds to the functional N_a used in Chapter 6 of Lukacs (1970) and has the following properties:

(i) $M_r(f) \geq 0$;
(ii) $M_r(f) = 0$ if and only if $f \equiv 1$;
(iii) If g is a proper component of f then $M_r(g) < M_r(f)$, provided $f'(0) = g'(0) = 0$.

Property (i) of the functional M_r follows immediately from $f(0) = 1$.[*] To derive (ii), we use the ridge property and Corollary 4 to Theorem 14.2.1. Let $g(z)$ be a proper ridge component of $f(z)$ so that $f(z) = g(z) k(z)$, where $f'(0) = g'(0) = 0$. It is then easily seen that $k'(0) = 0$. Since $k(iy) > 1$ we have $f(iy) > g(iy)$, so that (iii) is proved.

We prove here only Theorem 14.3.1. The proof of Theorem 14.3.2 is similar to the proof of Theorem 6.2.1, and we refer the reader to Tupicina (1972).

Proof of Theorem 14.3.1. We give an indirect proof and consider a ridge function $f(z)$ which has no indecomposable ridge component. In contradiction with the theorem, we assume that there exists a point z_0 in S_R such that

---

[*] We recall that we have assumed in Section 14.3 that all ridge functions considered are standardized.

$f(z_0) = 0$. Let A be the set of all ridge components of $f(z)$ which vanish at z_0. Clearly, $A \ne \emptyset$ since $f(z) \in A$. We write $\tilde{g}(z)$ for any ridge function such that $\tilde{g}'(0) = 0$. According to Lemma 14.3.1, one can find for each ridge function g a ridge function $\tilde{g}$ which has this property. Let

$$(14.3.7) \quad \gamma = \inf_{g \in A} M_r(\tilde{g}).$$

Let $\{g_n\}$ be a sequence of ridge components of f such that $M_r(\tilde{g}_n)$ tends to γ as $n \to \infty$. We see from Lemma 14.3.2 that the functions $\tilde{g}_n$ are uniformly bounded in the strip S_1. We select from the sequence $\{\tilde{g}_n\}$ a subsequence $\{\tilde{g}_{n_k}\}$ which is uniformly convergent in S_1. Let

$$g(z) = \lim_{k \to \infty} \tilde{g}_{n_k}(z).$$

According to Lemma 14.3.3 $g(z)$ is a ridge component of $f(z)$ such that $g(z_0) = 0$; hence $g(z) \in A$. Moreover $M_r(g) = \gamma$. It follows from $\tilde{g}'_n(0) = 0$ that $g'(0) = 0$. The ridge function $g(z)$ cannot have the form $\exp(iaz)$ since $g(z_0) = 0$. Hence we conclude from property (ii) of the functional M_r that $\gamma > 0$. Since the ridge function $f(z)$ has no indecomposable factor, its ridge component $g(z)$ is decomposable. Let

$$g(z) = k_1(z) k_2(z)$$

be a decomposition of $g(z)$ where none of the factors is degenerate. Then at least one of these factors must vanish at the point z_0. Let this be $k_1(z)$, then $k'_1 \in A$. According to Lemma 14.3.1 we can then assume that $k'_1(0) = 0$. Since $k_1(z) \ne g(z)$, we conclude from property (iii) of the functional M_r that $M_r(k_1) < M_r(g)$. Therefore $M_r(k_1) < \gamma$, which is impossible on account of the definition of γ, so that the proof of Theorem 14.3.1 is completed.

Corollary to Theorem 14.3.1. There exist indecomposable ridge functions. Every ridge function which has at least one zero in S has an indecomposable ridge component.

Example. The ridge function $f(z) = \cos t$.

Tupicina (1972) gave an example of an entire ridge function which is indecomposable. For the construction of this example we need the following lemma which is also of independent interest.

Lemma 14.3.4. The function $f_\alpha(z) = (1 - z^2) e^{-\alpha z^2}$ ($\alpha > 0$) is a ridge function for $\alpha \geqslant \theta$ but is not a ridge function for $0 < \alpha < \theta$. Here θ is the zero of the equation $1 + \theta + \log \theta = 0$.

This lemma is due to Ostrovskii who communicated it with its proof to Tupicina. Lemma 14.3.4 was published in Tupicina's paper. It can be shown that $f_\theta(z)$ is indecomposable and that $0.27 < \theta < 0.28$.

RIDGE FUNCTIONS

Remark 1. The characteristic function of an indecomposable distribution can be a decomposable ridge function.

Consider the function $f_\alpha(z) = (1-z^2)\,e^{-\alpha z^2}$ of Lemma 14.3.4 for $\alpha = \tfrac{1}{2}$. It is known (see Section 6.3 of Lukacs (1970), p. 184) that $f_{1/2}(z)$ is an indecomposable characteristic function. As a characteristic function it is necessarily a ridge function which admits the decomposition

$$f_{1/2}(z) = f_\theta(z)\exp[-(\tfrac{1}{2}-\theta)z^2].$$

The factor $f_\theta(z)$ is a ridge function but not a characteristic function.

Remark 2. A ridge function $f(z)$ may have a factor $g(z)$ whose strip of regularity is a proper subset of the strip of regularity of $f(z)$.

This is again a property which analytic characteristic functions do not share with ridge functions.

Let $\alpha \geqslant \tfrac{1}{2}$, $z = t + iy$ (t, y real) and

$$g_\alpha(z) = (1-iz)\,e^{-\alpha z^2}.$$

A simple computation shows that the entire function $g_\alpha(z)$ has the ridge property, provided that

$$\alpha \geqslant \frac{1}{2t^2}\log\left[1 + \frac{t^2}{(1+y)^2}\right].$$

If we maximize the right-hand side of this inequality we see that $g_\alpha(z)$ has the ridge property in the strip $|\operatorname{Im} z| < 1 - \sqrt{\dfrac{1}{2\alpha}}$. We also see easily that

$$h(z) = \frac{1}{1-iz}$$

is a ridge function in the half-plane $\operatorname{Im} z > -i$. The function $g_\alpha(z)\,h(z) = e^{-\alpha z^2}$ is an entire ridge function but its ridge factor $h(z)$ is not an entire function.

We note that the ridge function $h(z)$ can be continued analytically beyond its strip of regularity, but that the ridge property would not be preserved outside its strip.

14.4 Entire ridge functions

In this section we deal primarily with the growth of entire ridge functions. We mention first some results which entire ridge functions share with entire characteristic functions.

The order of a non-degenerate ridge function is at least equal to 1 [see Theorem 7.1.3 of Lukacs (1970)]. For any $\rho \geqslant 1$ one can construct an entire ridge function of order ρ. An analogue to Marcinkiewicz' result (Theorem 7.3.4 of Lukacs (1970)) is also valid.

Theorem 14.4.1. A non-constant entire function of finite order $\rho > 2$ whose exponent of convergence ρ_1 is less than ρ cannot be a ridge function.

Corollary 1. If an entire ridge function of finite order ρ has an exponent of convergence $\rho_1 < \rho$ then necessarily $\rho \leq 2$.

Corollary 2. An entire ridge function of finite order without zeros necessarily has the form $f(z) = k \exp(\alpha z + \beta z^2)$ (k complex, α, β real, $\beta \geq 0$).

The validity of these statements is not surprising since the use of the ridge property was the main tool in proving Marcinkiewicz' theorem. We have already mentioned Ostrovskii's (1963) generalizations of Marcinkiewicz' theorem which are applicable to classes of entire functions satisfying conditions (7.3.45a) or (7.3.45b) of the author's (1970) book. These classes contain the ridge functions, so that Ostrovskii's interesting results also hold for ridge functions.

Goldberg & Ostrovskii (1974) have studied entire ridge functions of finite order which have only real zeros. They obtained the following result:

Theorem 14.4.2. Let $f(z)$ be an entire ridge function of finite order which has only real zeros. Then it has the form

$$f(z) = c \exp[-\gamma z^2 + i\beta z] \prod_k \left(1 - \frac{z^2}{a_k^2}\right),$$

where c, α, β and a_k are constants, $\gamma \geq 0$, β real, $a_k > 0$, $\sum_k a_k^2 < \infty$.

Ostrovskii (1963) studied the growth of entire characteristic functions and obtained results which are applicable to ridge functions. Similar questions were also treated by Kaminin & Ostrovskii (1975). Some related work was done by V. V. Zimogljad (1969) who obtained the following result:

Theorem 14.4.3. Let $P(z)$ and $Q(z)$ be entire functions such that $P(z)$ is not a constant, and suppose that $M(r, P) = |P(z)|$ for $0 < r < \infty$. Assume that the function

$$f(z) = P\{\lambda_1 e^{iz} + \lambda_2 e^{-iz} + Q(z)\} \qquad (\lambda_1 \geq 0, \ \lambda_2 \geq 0)$$

is a ridge function. Then $Q(z)$ is either a polynomial of at most second degree or it is an entire function of order greater than or equal to 1.

The last-mentioned four papers use rather special methods from the theory of entire functions. A detailed discussion would exceed the scope of this monograph.

We next prove a smoothing theorem for the factors of a ridge function.

Theorem 14.4.4. Let $f(z)$ be a standardized ridge function and suppose that $f_1(z)$ is a standardized ridge factor of $f(z)$ and that $f(z)$ is regular in the strip

RIDGE FUNCTIONS

$S = \{z: a < \operatorname{Im} z < b\}$, where $a \leq 0 \leq b$ and $z = t + iy$. Then

(14.4.1) $\quad 1 \leq \left| \dfrac{f_1(iy)}{f_1(t+iy)} \right| \leq \left| \dfrac{f(iy)}{f(t+iy)} \right|$

in S.

Proof. Since f_1 is a ridge factor of f, there exists a ridge factor f_2 such that $f(z) = f_1(z) f_2(z)$. Therefore

(14.4.2) $\quad \dfrac{f(iy)}{f(t+iy)} = \dfrac{f_1(iy)}{f_1(t+iy)} \dfrac{f_2(iy)}{f_2(t+iy)}$.

Since $f_2(z)$ is a ridge function one has $|f_2(iy)/f_2(t+iy)| \geq 1$, and the inequality on the right-hand side of (14.4.1) follows from (14.4.2). The inequality on the left-hand side of (14.4.1) is the ridge property for $f_1(z)$.

15 Functions of bounded variation

In this chapter we study another generalization of characteristic functions by considering functions of bounded variation[*] instead of distribution functions. We shall discuss analytical properties of Fourier-Stieltjes transforms of this class of functions and we shall extend the theorems of Cramér and Raikov.

Let the function $f(x)$ be defined in the finite interval $[a, b]$. Consider the partition

$$D: a = t_0 < t_1 < t_2 < \ldots < t_n = b$$

of this interval and let

$$S_D(f) = \sum_{k=1}^{n} |f(t_k) - f(t_{k-1})|.$$

If there exists a finite number M such that $S_D(f) \leq M$ for all possible partitions of $[a, b]$ then $f(x)$ is said to be a function of bounded variation in $[a, b]$. The supremum of the set of all $S_D(f)$ is called the total variation of f in $[a, b]$ and is denoted by $V_a^b[f]$.

A complex-valued function $f(x)$ is said to be of bounded variation if its real as well as its imaginary part is of bounded variation.

A function $f(x)$ defined on $(-\infty, +\infty)$ is said to be a function of bounded variation if the numbers $V_a^b[f]$ form a bounded set for all intervals $[a, b]$. Then

$$\lim_{\substack{b \to \infty \\ a \to -\infty}} V_a^b[f] = V_{-\infty}^{\infty}[f]$$

is called the total variation of f over $(-\infty, +\infty)$. The variation in the intervals $(-\infty, 0)$ or $(0, +\infty)$ are defined in a similar manner.

We also write

$$\underset{|x|>y}{V}[f(x)] \quad \text{for} \quad V_{-\infty}^{-y}[f(x)] + V_y^{\infty}[f(x)].$$

15.1 Fourier-Stieltjes transforms of functions of bounded variation

Cramér (1939) derived a necessary and sufficient condition which ensures that a function is the Fourier transform of a function of bounded variation.

---

[*] For a discussion of properties of functions of bounded variation see for instance E. Hille (1964), vol. 1, Appendix C.

FUNCTIONS OF BOUNDED VARIATION

In order to obtain theorems which are analogous to results on characteristic functions one must restrict the consideration to functions of bounded variation $W(x)$ such that $W(-\infty) = 0$ while $W(+\infty) = 1$. We call the class of these functions the class B. It will be necessary to impose additional restrictions on functions of class B.

We state several properties of functions of bounded variation.

(I) Every function of bounded variation can be represented as the difference of two bounded non-decreasing functions. The converse of this statement is also true. Therefore one can write a function of class B as $W(x) = a_1 F_1(x) - a_2 F_2(x)$, where F_1 and F_2 are distribution functions and $a_1 - a_2 = 1$.

(II) Functions of bounded variation have only discontinuities of the first kind. The set of discontinuity points is enumerable.

(III) If α is a constant then $V_a^b[\alpha f] = |\alpha| V_a^b[f]$.

(IV) If f and g are functions of bounded variation then

$$V_a^b[f+g] \leqslant V_a^b[f] + V_a^b[g].$$

(V) If $a < b < c$ and if f is of bounded variation then

$$V_a^b[f] + V_b^c[f] = V_a^c[f].$$

(VI) Let W be a function of bounded variation. As a consequence of property (I), the Fourier-Stieltjes transform $w(t) = \int_{-\infty}^{\infty} e^{itx} \, dW(x)$ exists for all real t.

(VII) Let $w(t)$ be the Fourier-Stieltjes transform of the function $W(x)$ of bounded variation. For every pair x_1 and x_2 of continuity points of $W(x)$ one has

$$W(x_2) - W(x_1) = \frac{1}{2\pi} \lim_{T \to \infty} \int_{-T}^{T} \frac{e^{-itx_1} - e^{-itx_2}}{it} w(t) \, dt.$$

Yu. P. Studnev (1967, 1970) studied limit theorems in class B and introduced infinitely divisible functions of class B in the second paper.

(VIII) It is possible to define the convolution of two functions W_1 and W_2 from class B as

$$W(x) = \int_{-\infty}^{\infty} W_1(x-y) \, dW_2(y) = \int_{-\infty}^{\infty} W_2(x-y) \, dW_1(y)$$

and to show that $W(x) \in B$. Let $w_j(t)$ be the Fourier-Stieltjes transform of $W_j(x)$ and let $w(t)$ be the Fourier-Stieltjes transform of $W(x)$; then

$$w(t) = w_1(t) w_2(t),$$

so that the convolution theorem holds for the functions of bounded variation.

Fourier-Stieltjes transforms of functions of class B can be analytic or entire functions. However, an analogue to Theorem 8.1.1 of Lukacs (1970) does not hold.(*) This is shown by the following example which is due to R. G. Laha (1964). Let

(15.1.1a) $w(t) = e^{-t^2}$

(15.1.1b) $w_1(t) = (1+t^2) e^{-t^2/2}$

(15.1.1c) $w_2(t) = (1+t^2)^{-1} e^{-t^2/2}$.

The functions $w(t)$ and $w_2(t)$ are characteristic functions, so that $W(x)$ as well as $W_2(x)$ belong to B. A simple computation shows that $w_1(t)$ is the transform of

$$W_1(x) = \frac{1}{2\pi} \int_{-\infty}^{x} (2-y^2) e^{-y^2/2} \, dy.$$

Hence $W_1(x) \in B$. Clearly, $w(t) = w_1(t) w_2(t)$. The functions $w(t), w_1(t), w_2(t)$ can be continued analytically and are defined for complex values of their argument $z = t + iy$. We see that $w(z)$ and $w_1(z)$ are entire functions, while $w_2(z)$ has poles at $+i$ and $-i$ and is therefore regular only in the strip $|\text{Im}(z)| < 1$.

Let B_0 be a class of functions of bounded variation in $(-\infty, +\infty)$ which satisfies the following conditions:

(i) Every $W(x) \in B_0$ has at least two points of increase;

(ii) $\int_{-\infty}^{\infty} e^{iux} |dW(x)| < \infty$ for all real u;

(iii) The Fourier-Stieltjes transform $w(t)$ of $W(x)$ has the form $w(t) = \exp[P(t)]$, where $P(t)$ is a polynomial of degree greater than or equal to 2.

R. G. Laha (1964) proved the following statement:

Theorem 15.1.1. *Suppose that* $W(x) \in B_0$. *Then* $P(t)$ *is of even degree and has the form*

(15.1.2) $P(t) = \alpha_0 + i\alpha_1 t + \alpha_2 t^2 + i\alpha_3 t^3 + \ldots + \alpha_{2m} t^{2m},$

where the coefficients of $P(t)$ *are real and* $\alpha_{2m} < 0$.

Proof. We consider first the case where the degree of $P(t)$ is even, say $2m$. Then $w(z)$ is an entire function of z ($z = t + iy; t, y$ real) and

$$w(iy) = \int_{-\infty}^{\infty} e^{-yx} \, dW(x)$$

exists for all real y and is real. Then it follows that $P(t)$ has the form (15.1.2) with real coefficients. The boundedness of $w(t)$ implies $\alpha_{2m} < 0$.

(*) Theorem 8.1.1 of Lukacs (1970) states the following. Suppose that $f(z)$ is an analytic characteristic function which has $f_1(z)$ as a factor; then $f_1(z)$ is regular, at least in the strip of regularity of $f(z)$.

FUNCTIONS OF BOUNDED VARIATION   153

Next suppose that $P(t)$ has an odd degree, say $2m+1$. Using the above reasoning we see that

(15.1.3) $P(t) = \alpha_0 + i\alpha_1 t + \alpha_2 t^2 + \ldots + \alpha_{2m} t^{2m} + i\alpha_{2m+1} t^{2m+1}.$

Here all coefficients are real and $\alpha_{2m} < 0$. Let u be an arbitrary real number. The function

(15.1.4) $h(z) = w(z + iu) = \int_{-\infty}^{\infty} e^{izx - ux} \, dW(x)$

is then the Fourier-Stieltjes transform of

$$H(x) = \int_{-\infty}^{x} e^{-uy} \, dW(y),$$

where $H(x) \in B_0$. Therefore $h(t)$ is bounded for all real t and we see from (15.1.3) and (15.1.4) that

(15.1.5) $h(t) = w(t + iu) = \exp\{\alpha_0 + i\alpha_1(t + iu) + \ldots + i\alpha_{2m+1}(t + iu)^{2m+1}\}.$

The coefficient of t^{2m} in (15.1.5) is $\alpha_{2m} - (2m+1)u\alpha_{2m+1}$, and it can always be made positive by an appropriate selection of u; then $h(t)$ becomes unbounded for real t. This is impossible, so that the degree of $P(t)$ cannot be odd.

15.2 Decomposition of functions of bounded variation

R. G. Laha (1964) also obtained the following decomposition theorem for functions of B_0.

Theorem 15.2.1. *Suppose that the function $W(x)$ belongs to B_0 and that it admits the decomposition*

$$W(x) = W_1(x) * W_2(x)$$

where the "factors" $W_j(x)$ $(j = 1, 2)$ satisfy the conditions

(i) $W_j(x) \in B$

(ii) $\begin{cases} V_y^{\infty}[W_j(x)] = O[\exp(-y^{1+\delta})] \\ V_{-\infty}^{-y}[W_j(x)] = O[\exp(-y^{1+\delta})] \end{cases}$

as $y \to \infty$. Here $\delta > 0$ and $y > 1$. Then $W_j(x) \in B_0$.

In order to derive the theorem the following lemma is needed.

Lemma 15.2.1. *Suppose that a function $W(x) \in B_0$ satisfies the following conditions:*

(15.2.1) $\begin{cases} V_v^{\infty}[W(x)] = O[\exp(-v^{1+\delta})] \\ V_{-\infty}^{-v}[W(x)] = O[\exp(-v^{1+\delta})] \end{cases}$

as $v \to \infty$. Here δ is a fixed positive number and $y > 1$. Let $w(t)$ be the Fourier-Stieltjes transform of $W(x)$. Then the continuation $w(z)$ of $w(t)$, as a function of the complex variable z ($z = t + iy$; t, y real), is an entire function of some finite order $\rho \leqslant 1 + \delta^{-1}$.

For the proof of this lemma we refer the reader to R. G. Laha (1964).

We proceed to the proof of Theorem 15.2.1. Let $w_j(t)$ be the Fourier-Stieltjes transform of $W_j(x)$ ($j = 1, 2$). It follows from Lemma 15.2.1 that $w_1(z)$ and $w_2(z)$ are entire functions of finite order not exceeding $1 + 1/\delta$. Moreover,

$$w_1(z) w_2(z) = \exp[P(z)]$$

for all complex z. Therefore $w_1(z)$ and $w_2(z)$ are entire functions without zeros and, according to Hadamard's factorization theorem,

$$w_j(z) = \exp[P_j(z)] \qquad (j = 1, 2).$$

Here $P_j(z)$ is a polynomial of degree not exceeding $1 + 1/\delta$. Hence $W_1(x)$, as well as $W_2(x)$, belong to the class B_0. It follows finally from Theorem 15.1.1 that $P_j(t)$ is a polynomial of even degree, say $2m_j < \min(2m, 1 + 1/\delta)$, and has the form

$$P_j(t) = \alpha_{0,j} + i\alpha_{1,j} t + a_{2,j} t^2 + \ldots + \alpha_{2m,j} t^{2m_j}.$$

Here the coefficients are real and $\alpha_{2m,j} < 0$ for $j = 1, 2$. The relation

$$P_1(t) + P_2(t) = P(t)$$

is valid for all real t. Here $P(t)$ is the polynomial which occurs in the expression

$$W(t) = \exp[P(t)]$$

of $W(t) \in B_0$.

If $W(x)$ is a distribution function, one concludes from the theorem of Marcinkiewicz that $P(t)$ is a polynomial of second degree. This is also true for $P_1(t)$ and $P_2(t)$, so that W_1 and W_2 are normal distributions.

Corollary to Theorem 15.2.1. *If a normal distribution is the convolution of two functions of bounded variation satisfying the conditions of Theorem* 15.2.1 *then the factors are also normal distributions.*

This result is a generalization of Cramér's theorem. It is due to Yu. V. Linnik & V. P. Skitovič (1958). See also Linnik (1964), p. 101.

Theorem 15.2.2 (*Linnik-Skitovič*). *Suppose that the normal distribution $\Phi(x)$ admits a factorization*

$$\Phi(x) = W_1(x) * W_2(x),$$

where the $W_j(x)$ are symmetric functions of bounded variation such that

$$\int_{-\infty}^{\infty} dW_j(x) = 1$$

and where for some $\delta > 0$,

(15.2.2) $\quad V_{|x|>y} [W_j(x)] = O[\exp(-y^{1+\delta})] \quad$ as $y \to \infty$.

Then $W_1(x)$ and $W_2(x)$ are normal distribution functions.

In Linnik (1964) an example is given which shows that (15.2.2) cannot be replaced by the assumption that the variation of the W_j outside $(-y, y)$ is of order $O\left[\exp\left(-\frac{y}{2} \log(y+1)\right)\right]$ as $y \to \infty$.

G. P. Čistjakov (1970) generalized Theorem 15.2.2. For its formulation one has to define certain subsets of class B.

(1) Let $B_1 \subset B$ be the set of all functions W of bounded variation which satisfy

  (a) $W(x) + W(-x) = 2W(0) \quad$ for all x in $(-\infty, \infty)$;

  (b) $V_y^{\infty} [W(x)] = O[\exp(-y^{1+\delta})] \quad$ as $y \to \infty$ for some $\delta > 0$.

(2) Let $B_2 \subset B$ be the set of all functions W of bounded variation which admit the representation $W(x) = \omega(x) - \sigma(x)$, where $\omega(x)$ and $\sigma(x)$ are non-decreasing symmetric functions such that

$$V_y^{\infty} [\sigma(x)] = O[\exp(-y^{1+\delta})] \quad \text{as } y \to \infty \text{ for some } \delta > 0.$$

(3) Let B_3 be the set of all functions $W(x)$ of B_1 whose Fourier-Stieltjes transforms

$$w(t) = \int_{-\infty}^{\infty} e^{itx} dW(x)$$

can be continued analytically into some domain of the complex plane and for which $w(iy) \neq 0$.

Theorem 15.2.3. *Suppose that W_1 and W_2 belong to B_3. If $W_1 * W_2 = \Phi$ then W_1 and W_2 are normal distributions.*

Čistjakov also gives an example to show that it is not possible to replace the assumption $W_1 \in B_3$, $W_2 \in B_3$ by the weaker assumption $W_1 \in B_2$, $W_2 \in B_2$. Let

$$W_1(x) = \frac{1}{2} \int_{-\infty}^{x} e^{-|s|} ds, \qquad W_2(x) = \frac{1}{2\pi} \int_{-\infty}^{x} e^{-s^2/2}(2 - s^2) ds.$$

It is easily seen that $W_1, W_2 \in B_2$ and that

$$w_1(t) = \frac{1}{1+t^2}, \qquad w_2(t) = (1 + t^2) e^{-t^2/2},$$

so that $W_1 * W_2 = \Phi$. However, $w_2(i) = 0$, so that $W_2 \notin B_3$.

For the proof, an analogue to Theorem 1.4.1 is needed. For this we introduce a subclass $B_4 \subset B_2$ of function $W(x)$ such that their Fourier-Stieltjes transform $w(t)$ can be continued into a horizontal strip $\{|\operatorname{Im} z| < h\}$ and for which $w(iy) \neq 0$ for $|y| < h$.

Theorem 15.2.4. *Suppose that W_1 and W_2 are such that $W_1 * W_2 = W$. Here $W \in B_4$ with the strip $|\operatorname{Im} z| < h$. Then the Fourier-Stieltjes transforms of W_1 and W_2 are also regular in the strip.*

For the proof of Theorems 15.2.3 and 15.2.4 we refer the reader to Čistjakov's (1970) paper.

Čistjakov (1976) gave other extensions of Theorem 15.2.2 in which he modified and weakened assumption (15.2.2). We mention here only one of his results, and we introduce for this purpose another class of functions of bounded variation.

Let B_5 be the class of right continuous functions $W(x)$ of bounded variation such that

(a) $W(-\infty) = 0$; $W(+\infty) = 1$.
(b) $W(x)$ admits the representation $W(x) = \omega(x) - \sigma(x)$, where $\omega(x)$ and $\sigma(x)$ are non-decreasing and where for all $x > 0$,

$$\underset{|x|>y}{V} [\sigma(x)] = O[\exp(-cy \log y)] \quad \text{as } y \to \infty.$$

(c) The Fourier-Stieltjes transform $w(t) = \int_{-\infty}^{\infty} e^{itx} \, dW(x)$ has the property that $w(iy) \neq 0$.

Theorem 15.2.5. *Suppose that $W_1, W_2 \in B_5$ and that $W_1 * W_2 = \Phi$; then W_1 and W_2 are normal (possibly degenerate) distribution functions.*

The proof of Theorem 15.2.5 can be found in Čistjakov (1976).

In this section we have presented several extensions of Cramér's theorem. These assume that a normal distribution could be the convolution of two functions of a subclass of the functions of bounded variation, which were not supposed to be distribution functions. These theorems always require some conditions on the growth of the variation of the functions of bounded variation. These restrictions prevented the Fourier-Stieltjes transforms from being entire functions of infinite order. Therefore one could not obtain a similar generalization of Raikov's theorem. This situation motivated the introduction of a different condition for defining a new subclass of the functions of bounded variations.

Let $\mathscr{B}$ be the class of function $W(x)$ of bounded variation for which, for all real $x > 0$,

(15.2.3a) $\quad W(-\infty) = 0, \quad W(+\infty) = 1$

FUNCTIONS OF BOUNDED VARIATION 157

(15.2.3b) $\quad \underset{|y|>x}{V} [W(y)] = V_{-\infty}^{-x}[W(x)] + V_x^{\infty}[W(x)] = O(e^{-rx})$

hold.

The Fourier-Stieltjes transforms of functions $W \in \mathscr{B}$ are entire functions. This follows from property (II) of the functions of bounded variation and from Theorem 1.4.8.

We now introduce a new concept which will permit us to define a subset of $\mathscr{B}$ which we shall use below.

A real-valued function $f(x)$ is said to be ultimately positive if there exists an $x_0 < \infty$ such that $f(x) \geq 0$ for all $x \geq x_0$ and $f(x_1) > 0$ for at least one finite $x_1 > x_0$. We then write $f(x) \in U$ or $f(x) \in U(x_0)$ if we wish to specify x_0.

Let $\mathscr{B}_1 \subset \mathscr{B}$ be the set of functions $W(x)$ of bounded variation which satisfy (15.2.3a) and (15.2.3b) and for which $W(-x)$ as well as $1 - W(x)$ are ultimately positive.

Theorem 15.2.6. *Let $W \in \mathscr{B}_1$ and let w be the Fourier-Stieltjes transform of W. Then there exists a horizontal strip $S = \{z = t + iy: y_1 > y > y_2\}$ which contains the real axis, such that*

(15.2.4a) $\quad w(iy) > \exp(-|y|a)$

(15.2.4b) $\quad |w(z)| \leq A|z|w(iy)$

outside the strip S. Here a and A are positive constants.

For the proof we need several lemmas.

Lemma 15.2.1A. *If $W \in \mathscr{B}$ then the Fourier-Stieltjes transform of W can be written in the form*

(15.2.5a) $\quad w(z) = -iz \int_{-\infty}^{\infty} e^{izx} W(x) \, dx \qquad \text{if } \operatorname{Im}(z) > 0,$

but in the form

(15.2.5b) $\quad w(z) = iz \int_{-\infty}^{\infty} e^{izx} [1 - W(x)] \, dx \qquad \text{if } \operatorname{Im}(z) < 0.$

The lemma follows from a property of the Laplace transform [D. V. Widder (1946), p. 239] by a simple substitution.

Lemma 15.2.2. *Let $W(x) \in \mathscr{B}$ and suppose that $W(x) \in U(-x_0)$. Let $w(z)$ be the Fourier-Stieltjes transform of W, $[z = t + iy; t, y \text{ real}]$. Then there exists a $y_1, 0 < y_1 < \infty$, such that $y > y_1$ implies*

(15.2.6a) $\quad w(iy) > \exp(-yx_0)$

and

(15.2.6b) $\quad |w(z)| \leq A_1|z|w(iy)$

for some positive constant A_1 and x_0.

Proof. Assume $\text{Im}(z) = y > 0$. Then it follows from (15.2.5a) that

(15.2.7a) $\quad w(iy) = y \int_{-\infty}^{\infty} e^{-yx} W(x)\, dx$

or

(15.2.7b) $\quad w(iy) = y \int_{-\infty}^{x_0} e^{-yx} W(x)\, dx + y \int_{x_0}^{\infty} e^{-yx} W(x)\, dx.$

Let $x_0 < 0$; then if $x \in (-\infty, x_0)$ one has $-x > -x_0 > 0$ and $W(x) \geq 0$. Thus the first integral in (15.2.7b) is non-negative while the second integral could be negative. Therefore,

(15.2.8) $\quad |w(iy)| \geq y \int_{-\infty}^{x_0} e^{-yx} W(x)\, dx - y \left| \int_{x_0}^{\infty} e^{-yx} W(x)\, dx \right|.$

To sharpen this inequality we minimize the first term and maximize the second. It follows from the assumption that $W(-x) \in U(-x_0)$ that there exists an x_1, $-\infty < x_1 < x_0 < 0$ and a $\delta > 0$ such that $W(x_1) = \delta > 0$. Then there exists an $\epsilon > 0$ such that $W(x) \geq \delta/2$ for $x \in (x_1, x_1 + \epsilon)$. If we choose $\epsilon > 0$ sufficiently small, then $x_1 + \epsilon < x_0$, that is

$$x_0 - (x_1 + \epsilon) = \epsilon_1 > 0 \quad \text{and} \quad x_0 - \epsilon_1 = x_1 + \epsilon.$$

Thus we minimize the first integral on the right-hand side of (15.2.8) as follows.

$$y \int_{-\infty}^{x_0} e^{-yx} W(x)\, dx \geq y \int_{x_1}^{x_1+\epsilon} e^{-yx} W(x)\, dx \geq \frac{\delta}{2} [e^{-yx_1} - e^{-y(x_1+\epsilon)}]$$

$$= \frac{\delta}{2} e^{-y(x_0-\epsilon_1)}(e^{\epsilon y} - 1) = e^{-yx_0} \frac{\delta}{2} e^{y\epsilon_1}(e^{y\epsilon} - 1),$$

so that

(15.2.9) $\quad y \int_{-\infty}^{x_0} e^{-yx} W(x)\, dx \geq e^{-yx_0} \frac{\delta}{2} e^{y\epsilon_1}(e^{\epsilon y} - 1).$

Let $M = \max(|W(x)|)$ and let $x \in (x_0, \infty)$. Then we obtain for the second integral in (15.2.7b)

(15.2.10) $\quad y \left| \int_{x_0}^{\infty} e^{-yx} W(x)\, dx \right| \leq M e^{-yx_0}.$

We substitute (15.2.8) and (15.2.9) into (15.2.7b) and get

$$w(iy) \geq e^{-yx_0} \left[\frac{\delta}{2} e^{y\epsilon_1}(e^{y\epsilon} - 1) - M \right].$$

FUNCTIONS OF BOUNDED VARIATION 159

The last inequality implies statement (15.2.6a) of the lemma, since there exists a $y_1 < \infty$ such that $y > y_1$ implies that the coefficient of e^{-yx_0} is greater than 1.

We next prove formula (15.2.6b) of the lemma; we take x_0 and M as before and again assume that $y > 0$. Then (15.2.5a) implies

$$(15.2.11) \quad |w(z)| \leqslant |z| \int_{-\infty}^{\infty} e^{-yx} |W(x)| \, dx.$$

Then, using the fact that $W(x) = |W(x)|$ for $x \in (-\infty, x_0)$, we see that

$$\int_{-\infty}^{\infty} e^{-yx} |W(x)| \, dx = \int_{-\infty}^{\infty} e^{-yx} W(x) \, dx - \int_{x_0}^{\infty} e^{-yx} [W(x) - |W(x)|] \, dx,$$

where

$$\left| \int_{x_0}^{\infty} e^{-yx} [W(x) - |W(x)|] \, dx \right| \leqslant 2M \, e^{-yx_0}/y.$$

Using the last inequality, the preceding equation and (15.2.7a), we get

$$(15.2.12) \quad \int_{-\infty}^{\infty} e^{-yx} |W(x)| \, dx \leqslant \left| \int_{-\infty}^{\infty} e^{-yx} W(x) \, dx \right|$$

$$+ \int_{x_0}^{\infty} e^{-yx} [W(x) - |W(x)|] \, dx \leqslant [|w(iy)| + 2M \, e^{-yx_0}]/y.$$

We use (15.2.6a) to conclude that there exists a finite y_1 such that, for $y > y_1$,

$$w(iy) > e^{-yx_0} > 0.$$

Therefore

$$(15.2.13) \quad \int_{-\infty}^{\infty} e^{-yx} |W(x)| \, dx \leqslant w(iy)(1 + 2M) \, y_1^{-1}.$$

It follows finally from (15.2.11) and (15.2.13) that

$$|w(z)| \leqslant z w(iy)(1 + 2M) \, y_1^{-1} = A|z|w(iy)$$

for $y > y_1$. This is (15.2.6b), so that the lemma is proved.

Lemma 15.2.3. Suppose that $W(x) \in \mathcal{B}$ and that the Fourier-Stieltjes transform of $W(x)$ is $w(z)$ $(z = t + iy)$. Assume further that $[1 - W(x)] \in U(x_0')$. Then there exists a y_2, $-\infty < y_2 \leqslant 0$, such that $y < y_2$ implies

(a) $w(iy) > \exp(yx_0')$
(b) $|w(z)| \leqslant A_2 |z| w(iy)$

for some positive constant A_2.

The proof is effected in the same way as the proof of the preceding lemma except that (15.2.5b) is used instead of (15.2.5a).

Theorem 15.2.6 follows immediately from Lemmas 15.2.2 and 15.2.3 by taking $A = \max(A_1, A_2)$ and $a = \max(x_0, x_0')$.

We now prove an extension of Cramér's theorem[*] which uses the condition of ultimate positivity.

Theorem 15.2.7. Suppose that both W_1 and W_2 are in $\mathscr{B}_1$ and that

$$W_1 * W_2 = \frac{1}{\sqrt{2\pi}} \int_{-\infty}^{x} e^{-y^2/2} \, dy.$$

Then both W_1 and W_2 are normal distribution functions.

We conclude from the convolution theorem for functions of bounded variation and from the fact that the Fourier-Stieltjes transforms of functions from $\mathscr{B}$ are entire functions that

(15.2.14) $\quad w_1(z) w_2(z) = \exp(-z^2/2)$

for all complex z. Putting $z = iy$, we get

$$w_1(iy) w_2(iy) = \exp(y^2/2).$$

We apply (15.2.6a) to w_1 and see that there exists a constant x_0 and an interval I on the imaginary axis such that

$$w_1(iy) > \exp(-x_0|y|)$$

outside I; hence,

$$w_2(iy) < \exp(y^2/2) \exp(x_0|y|).$$

We see from (15.2.6b) that $|w_2(z)| \leq A_1|z|w_2(iy)$ and conclude that $w_2(z)$ is bounded. Moreover, $w_2(z)$ is an entire function of finite order $\rho \leq 2$, and we see from (15.2.14) that $w_2(z)$ has no zeros. It follows in the usual way that $w_2(t) = \exp(ict - dt^2)$ with c real and $d \geq 0$. The same reasoning applies to $w_1(t)$ so that the theorem is proved.

For the extension of Raikov's theorem[†] we need the following lemma.

Lemma 15.2.4. If $W(x) \in \mathscr{B}_1$ and if it is constant on $(-\infty, L]$ [respectively on $[R, \infty)$], and if L is the largest number [respectively R the smallest number] for which this holds, then

$$L = -\lim_{y \to \infty} [\log w(iy)/y] \; (\text{respectively } R = \lim_{y \to \infty} [\log w(-iy)/y]).$$

---

[*] Which states that a normal distribution has only normal factors.

[†] Raikov's theorem states that all factors of a Poisson distribution are Poisson distributions.

FUNCTIONS OF BOUNDED VARIATION

This is an extension of a theorem of G. Pólya (1949) in the slightly generalized formulation of B. Ramachandran (1962).

This lemma, as well as Theorems 15.2.7 and 15.2.8, are due to G. L. Grunkemeier (1975) where a proof of the lemma may be found.

Theorem 15.2.8. *Suppose that W_1 and W_2 are in $\mathscr{B}_1$ and that they are lattice-like with unit span,*(*) *and that $w_1(z) w_2(z) = \exp[\lambda(e^{iz} - 1)]$ with $\lambda > 0$. Then W_1 and W_2 are both Poisson distributions.*

It follows from the definition of W_1 and W_2 that for y real,

(15.2.15) $\quad w_1(iy) w_2(iy) = \exp[\lambda (e^{-y} - 1)].$

We take logarithms, divide by y and let y tend to infinity; we note that the right-hand side of (15.2.4) tends to zero, so that

$$\lim_{y \to \infty} \frac{\log w_1(iy)}{y} = - \lim_{y \to \infty} \frac{\log w_2(iy)}{y} = B,$$

where B is a constant. By Lemma 15.2.4 the functions W_1 and W_2 are bounded to the left by $-B$ and B respectively. Without loss of generality we may assume that $B = 0$. This can always be accomplished by multiplying the w_1 and w_2 by an unessential factor.

Thus we get

(15.2.16) $\quad w_1(z) = \sum_{k=0}^{\infty} a_k e^{izk}, \qquad w_2(z) = \sum_{k=0}^{\infty} b_k e^{izk}$

(15.2.17) $\quad W_1(x) = \sum_{k \leqslant x} a_k, \qquad W_2(x) = \sum_{k \leqslant x} b_k,$

where $a_k \gtreqless 0$, $b_k \gtreqless 0$. We introduce the new variable $v = e^{iz}$ into (15.2.16) and in this way transform $w_1(z)$ and $w_2(z)$ into

$$h_1(v) = \sum_{k=0}^{\infty} a_k v^k, \qquad h_2(v) = \sum_{k=0}^{\infty} b_k v^k$$

and see that

(15.2.18) $\quad h_1(v) h_2(v) = \exp[\lambda(v - 1)].$

It follows from the definition of the class $\mathscr{B}_1$ that for any $r > 0$,

$$\bigvee_{|y| > N} [W_1(y)] = O(e^{-rN}) \qquad \text{as } N \to \infty$$

(*) A function W of bounded variation is said to be "lattice-like with unit span" if its only points of increase are the non-negative integers. No condition is imposed on the sign of the jumps at the lattice points, so that these functions are not necessarily distribution functions.

while

$$|a_n| \leq \sum_{k=N}^{\infty} |a_k| = \underset{|y|>N-1}{V} [W_1(y)]$$

so that

$$|a_N| = O[\exp(-r(N-1))] \quad \text{as } N \to \infty.$$

Therefore there exists a finite M ($M > 1$) and a positive integer N_0 such that

(15.2.19) $\quad |a_N|^{1/N} < M^{1/N} \exp\left(\frac{r}{N} - r\right)$

for $N > N_0$.

Let $\epsilon > 0$ and select r_0 so that $\exp(r_0) > (M^{1/N_0} e)/\epsilon$ $\quad |a_N|^{1/N} < \epsilon$ for $N > \max(r_0, N_0)$; hence

$$\lim_{N \to \infty} |a_N|^{1/N} = 0,$$

so that $h_1(v)$ is an entire function which has, according to (15.2.18), no zeros. The same reasoning applies to $h_2(v)$, so that $h_1(v)$ and $h_2(v)$ are entire functions without zeros.

We now study the order of these functions. We had $v = e^{iz}$, where $z = t + iy$. If we write $v = r e^{i\theta}$ then $r = e^{-y}$ and $t = \theta$; hence

$$w_j(iy) = h_j(r).$$

We see from Lemma 15.2.3 that there exists an r_0, $1 < r_0 < \infty$ such that for $r > r_0$

(15.2.20) $\quad \begin{cases} h_j(r) > r^{-a_j} \\ |h_j(z)| \leq A_j |\log z| h_j(r), \end{cases}$

$j = 1, 2$.

Here a_1 and a_2 and A_1, A_2 are positive constants. According to (15.2.18) one has $h_1(r) h_2(r) = \exp[\lambda(r-1)]$, and using (15.2.20) one sees that

$$|h_1(z)| \leq A_1 |\log z| \exp[\lambda(r-1)] r^{-a_2}.$$

Therefore $h_1(z)$ has at most order 1 and, by the same reasoning, the order of h_2 is also at most 1, and we conclude from Lemma 15.2.1 that these functions are exactly of order 1. It follows from Hadamard's factorization theorem and from the connection between the functions w_j and h_j that

$$w_j(z) = \exp[\mu_j(e^{iz} - 1)].$$

Finally we can easily show that $\mu_j > 0$ ($j = 1, 2$).

The class of ultimately positive functions of bounded variation does not include all distribution functions, so that Theorems 15.2.7 and 15.2.8 do not imply the theorem of Cramér or the theorem of Raikov, since both of these are valid for all distribution functions. N. I. Yakovleva (1978) has defined a new subclass of functions of bounded variation which contains the distribution functions, and she has derived a decomposition theorem analogous to Linnik's theorem on the factorization of a convolution of a normal and a Poissonian distribution. Her result therefore includes the theorems of Cramér and of Raikov.

Appendix

A1 Entire functions and their growth

A single-valued function $f(z)$ which is analytic in every finite region of the z-plane is called an entire function.

An entire function admits a power series expansion

$$(1.1) \quad f(z) = \sum_{k=0}^{\infty} c_k z^k$$

which is convergent for all finite z, so that

$$\lim_{n \to \infty} [a_n]^{1/n} = 0.$$

Let $f(z)$ be an entire function and suppose that there exists a number m such that $f(z)/z^m < M$ (M fixed); then $f(z)$ is a polynomial of degree at most equal to m. In this case $f(z)$ is at infinity either regular or has a pole. We are not interested in this case, but we shall consider only entire functions which have an essential singularity at infinity. Such functions are called entire transcendental functions. We denote by

$$(1.2) \quad M(r;f) = \max_{|z| \leqslant r} |f(z)|$$

the maximum modulus of $|f(z)|$ in the circle $|z| \leqslant r$. This value is assumed on the perimeter of the circle.

The order ρ of an entire function $f(z)$ is defined as

$$(1.3) \quad \rho = \limsup_{r \to \infty} \frac{\log \log M(r;f)}{\log r}.$$

One has $0 \leqslant \rho \leqslant \infty$. In this monograph we are in general not interested in functions of order inferior to 1; if $\rho < \infty$ then we say that $f(z)$ is an entire function of finite order.

An entire function $f(z)$ of finite order ρ is said to be of type τ if

$$(1.4) \quad \limsup_{r \to \infty} \frac{\log M(r;f)}{r^\rho} = \tau.$$

An entire function $f(z)$ of finite order ρ is said to be of minimal type if $\tau = 0$, of normal (or intermediate) type if $0 < \tau < \infty$, and of maximal type if $\tau = \infty$.

APPENDIX

Entire functions of order 1 and finite type ($\tau < \infty$) or of order inferior to 1 are called entire functions of exponential type.

Order and type of an entire function can be expressed in terms of its coefficients; one has

(1.5) $\quad \rho = \limsup\limits_{k \to \infty} \dfrac{k \log k}{\log |c_k|^{-1}}$

and, if $0 < \rho < \infty$,

(1.6) $\quad \tau = \dfrac{1}{e\rho} \limsup\limits_{k \to \infty} k|c_k|^{\rho/k}.$

For the proof of these statements we refer the reader to E. Hille (1962), pp. 182–8, or A. I. Markushevič (1965, see vol. II, ch. 9).

Order and type of entire functions describe their growth. One can also introduce, following G. Valiron (1949, 1959) and Levin, B. Ya. (1964), proximate orders and their types.

A real-valued function $\rho(r)$ is called a proximate order if it is defined for all $r > 0$, if it admits everywhere a left and right derivative, and if it satisfies the following conditions:

(i) $\lim\limits_{r \to \infty} \rho(r) = \rho \quad (0 < \rho < \infty)$ (ρ is the order of f);
(ii) $\rho'(r)$ exists for sufficiently large r;
(iii) $\lim\limits_{r \to \infty} \rho'(r) \, r \log r = 0.$

Let $f(z)$ be an entire function of order ρ; denote its maximum modulus by $M(r, f) = M(r)$ and suppose that

(iv) $\limsup\limits_{r \to \infty} \log M(r)/r^{\rho(r)} = \tau \quad (0 < \tau < \infty).$

Then $\rho(r)$ is called a proximate order of $f(z)$ and τ is called the type of $f(z)$ with respect to the proximate order $\rho(r)$.

Remark. The proximate order and type are not uniquely determined by $f(z)$: for example, if one adds $\log c/\log r$ to a proximate order of $f(z)$, then one obtains a new proximate order for $f(z)$, and the type has been multiplied by c.

We mention a few properties of proximate orders.

(1) The function $P(r) = r^{\rho(r)}$ is monotone increasing for sufficiently large r [see Levin (1964), p. 33]. Moreover,

$$\lim\limits_{r \to \infty} [P(kr)/P(r)] = k^\rho$$

uniformly in every interval $0 < a \leq k \leq b < \infty$.

(2) Let $\rho(x)$ be a proximate order and denote by $x = \phi(r)$ the inverse

function to $r = x^{\rho(r)}$. Then

$$\bar{\rho}(r) = \frac{1}{\rho[\phi(r)]}$$

is also a proximate order and is called the dual proximate order to the proximate order $\rho(r)$. One has

(1.7) $\quad r^{\bar{\rho}(r)} = \phi(r)$

and

$$\lim_{r \to \infty} \bar{\rho}(r) = \frac{1}{\rho}.$$

(3) It is easily seen that $\bar{\bar{\rho}}(r) = \rho(r)$.

(4) Let $\rho(x)$ be a proximate order, then

$$H(x) = \begin{cases} 0 & \text{if } x \leq x_0 \\ 1 - \exp(1 - x^{1+\rho(x)}) & \text{if } x > x_0 \end{cases}$$

is a distribution function, provided x_0 is chosen sufficiently large.

(5) Let $f(r, u) = e^{ru}[1 - H(u)]$. For all $r \geq 0$ there exists a value $u(r)$ such that the following two conditions are satisfied:

(i) $f(r, u(r)) \geq f(r, u) \quad$ for $u \leq u(r)$
(ii) $f(r, u(r)) > f(r, u) \quad$ for $u > u(r)$.

The function $u(r)$ is right continuous, non-decreasing, and satisfies the relation

$$\lim_{r \to \infty} u(r) = \infty.$$

For proofs we refer the reader to M. Dewess and M. Riedel (1977), H. J. Rossberg (1967) and B. Ya. Levin (1964).

It is sometimes useful to define some other measure of the growth of an entire function.

Let $f(z)$ be an entire function with maximum modulus $M(r)$. Then

$$\lambda = \liminf_{r \to \infty} \frac{\log \log M(r)}{\log r}$$

is called the lower order of $f(z)$.

We also present some results from the theory of entire functions which are needed.

Let $z_1, z_2, \ldots, z_w$, be a sequence of complex numbers which has no finite limit points and which is indexed in such a way that the moduli $r_k = |z_k|$ are non-decreasing. The exponent of convergence of the sequence $\{z_k\}_1^\infty$ is defined

APPENDIX

as the greatest lower bound of all $\lambda > 0$ for which the series

$$\sum_{z_k \neq 0} r_k^{-\lambda}$$

is convergent.

Theorem A1. Let $f(z)$ be an entire function of order ρ and suppose that the exponent of convergence of the zeros of $f(z)$ is ρ_1. Then $\rho_1 \leqslant \rho$.

Let

$$(1.8) \quad E(z;q) = \begin{cases} (1-z) & \text{if } q = 0, \\ (1-z)\exp\left(z + \dfrac{z^2}{2} + \ldots + \dfrac{z^q}{q}\right) & \text{if } q > 0. \end{cases}$$

The functions $E(z;q)$ are known as the Weierstrass primary factors.

Theorem A2. The product $Q(z;q) = \prod_k E\left(\dfrac{z}{z_k};q\right)$ is absolutely and uniformly convergent in every finite disk and $Q(z;q)$ is an entire function whose zeros are the points z_k. $Q(z;q)$ is called the Weierstrass canonical product of genus q. The order of a canonical product is equal to the exponent of convergence of its zeros.

Theorem A3 (Hadamard's factorization theorem). Each entire characteristic function $f(z)$ of finite order ρ admits a representation

$$f(z) = z^m \, e^{P(z)} \, Q(z;\rho),$$

where m is a non-negative integer, $P(z)$ a polynomial of degree not exceeding ρ, and where $Q(z;\rho)$ is a Weierstrass canonical product of genus $q \leqslant \rho$. (If $f(z)$ has no zeros then one takes $Q(z;\rho) \equiv 1$.)

For proofs of these results see E. Hille (1962) or A. I. Markuševič (1965), vol. 2.

Theorem A4. Let $H(z)$ be an entire function which has only real zeros. Denote the multiplicity of the zero of $H(z)$ at $z = 0$ by m ($m \geqslant 0$). Suppose that the order $H(z)$ is $\rho \leqslant 2$ while the exponent of convergence of the zeros of $H(z)$ is $\rho_1 < 2$. Assume further that $H(z)/z^m$ is an even function of z and that $H'(t)/t^m$ is bounded for real t. Then $H'(z)$ has only real zeros and $H'(z)$ has also order ρ and exponent of convergence ρ_1. Moreover, $H'(z)$ has a zero of order $m-1$ at $z = 0$ if $m \geqslant 1$, while it has a simple zero at $z = 0$ if $m = 0$.

A2 Schwarz's reflection principle

Let D_1 and D_2 be two domains such that $D_1 \cap D_2 = \emptyset$ while $\bar{D}_1 \cap \bar{D}_2$ is an interval γ on the real axis. Let $f_1(z)$ be regular in D_1 and continuous in $D_1 \cup \gamma$

and let $f_2(z)$ be regular in D_2 and continuous in $D_2 \cup \gamma$. Suppose that for $\xi \in \gamma$

$$\lim_{z \to \xi} f_1(z) = \lim_{z \to \xi} f_2(z) = h(\xi),$$

where the approach is from D_1 in the first and from D_2 in the second limit. Then there exists a function $f(z)$ regular in $D_1 \cup D_2 \cup \gamma$ which coincides with $f_1(z)$ in D_1 and with $f_2(z)$ in D_2.

For the proof we refer to Hille (1959), p. 184. In Chapter 11 of Lukacs (1970) we used for D_1 the interior of a rectangle located in the upper half-plane and for D_2 the interior of the rectangle which is located symmetrically to D_1 with respect to the real axis.

List of examples

Chapter and section	Page	Description of example
5.2	50	A unimodal F such that $F * F$ is not unimodal.
5.2	53	Convolution of a symmetric and an α-unimodal distribution can be β-unimodal with different values of β depending on α.
6.3	70	Factors of the Cauchy distribution (Theorem 6.3.1).
6.3	71	Factorization of symmetric stable laws (Theorem 6.3.2).
6.4	73	Absolutely continuous indecomposable distributions.
7.4	92	Characteristic functions in the van Dantzig class.
7.4	100	A characteristic function which is in $\mathscr{L}$ but not in I_0.
9.1	112	An indecomposable distribution may admit an α-decomposition.
10	118	A distribution which has a boundary characteristic function.
11.2	121	A mixture of exponential distributions which is not infinitely divisible.
13.1	132	Two metrics which are not comparable.
13.3	137	Existence of two distributions F_1 and F_2 with characteristic functions f_1 and f_2 such that $\lambda(f_1, f_2)$ is not small while $L(F_1, F_2)$ is small.
14.1	140	A ridge function which is not a characteristic function.
14.1	140	An entire ridge function which is not a characteristic function.
14.3	147	An indecomposable characteristic function can be a decomposable ridge function.
14.3	147	A ridge function $f(z)$ which has a factor whose strip of regularity is a proper subset of the strip of $f(z)$.
14.3	147	An entire ridge function which has a non-entire ridge factor.
15.1	152	An entire function of bounded variation can have a factor which is not an entire function.

References*

ABRAMOV, V. A. (1976). Estimates for the Lévy-Prohorov distance. *Theory Probab. Applic.*, **21**, 406-10.

ALF, C. & O'CONNOR, T. (1977). Unimodality of the Lévy spectral function. *Pacific J. Math.*, **69**, 285-90.

ASKEY, R. (1975). Some characteristic functions of unimodal distributions. *J. Math. Analysis Applic.*, **50**, 465-9.

BARNDORFF-NIELSEN, O. & HALGREEN, C. (1977). Infinite divisibility of the hyperbolic and generalized inverse Gaussian distributions. *Zeitschr. Wahrsch. verw. Gebiete*, **38**, 309-31.

BARTHOLOMEW, D. J. (1969). Sufficient conditions for a mixture of exponentials to be a probability density. *Ann. Math. Statist.*, **40**, 2183-8.

BERMAN, S. M. (1975). New characterization of characteristic functions of absolutely continuous distributions. *Pacific J. Math.*, **58**, 323-9.

BOAS, R. P. (1954). *Entire Functions.* Academic Press, New York.

BOAS, R. P. & SMITHIES, F. (1938). On the characterization of a distribution function by its Fourier transform. *Amer. J. Math.*, **60**, 523-31.

BOCHNER, S. (1932). *Fouriersche Integrale.* Akademische Verlagsgesellschaft, Leipzig. Reprinted by Chelsea Publ. Co., New York, 1948.

BONDESSON, L. (1979a). On the infinite divisibility of powers of Gamma variables. *Zeitschr. Wahrsch. verw. Gebiete*, **49**, 171-5.

BONDESSON, L. (1979b). On generalized Gamma and generalized negative binomial convolutions: I, II. *Skand. Aktuartidskr.*, 125-46, 147-66.

CHUNG, K. L. (1953). Sur les lois de probabilité unimodales. *C.R. Acad. Sci. Paris*, **236**, 583-4.

ČISTJAKOV, G. P. (1970). A generalization of the theorems of H. Cramér and of Yu. V. Linnik and V. P. Skitovič. *Theory Probab. Applic.*, **15**, 343-50.

ČISTJAKOV, G. P. (1971). On the conditions under which probability laws with non-analytic characteristic functions belong to class I_0. *Soviet Mat. Dokl.*, **12**, 1654-8.

ČISTJAKOV, G. P. (1975). On a problem of D. Dugué. *Soviet Mat. Dokl.*, **16**, 1315-19. Russian original in *Dokl. Akad. Nauk*, **224** (1975), No. 4.

ČISTJAKOV, G. P. (1976). Decomposition of a normal law into a convolution of functions of bounded variation. *Math. Notes*, **19**, 80-7. Russian original in *Mat. Zametki*, **19**, No. 1, 133-47.

CRAMÉR, H. (1939). On the representation of a function by certain Fourier integrals. *Trans. Amer. Math. Soc.*, **46**, 191-201.

CUPPENS, R. (1968). Sur un théorème de Paul Lévy. *Publ. Inst. Statist. Univ. Paris*, **XVII**, 1-6.

CUPPENS, R. (1975). *Decomposition of Multivariate Probabilities.* Academic Press, New York.

---

* Details are given in Cyrillic script of Russian language papers which had not appeared in English translation at the time of writing.

REFERENCES

DEWESS, M. & RIEDEL, M. (1977). The connection between the proximate order of an entire characteristic function and the corresponding distribution function. *Czechoslovak Math. J.*, **27**, **102**, 173-85.

DHARMADHIKARI, S. W. & JOGDEO, K. (1974). On characterizations of the unimodality of discrete distributions. *Ann. Inst. Statistical Math.*, **28**, 9-18.

DUGUÉ, D. (1951a). Analycité et convexité des fonctions caractéristiques. *Ann. Inst. Henri Poincaré*, **12**, 45-56.

DUGUÉ, D. (1951b). Sur certains exemples de décomposition des lois de probabilités. *Ann. Inst. Henri Poincaré*, **12**, 159-69.

DUGUÉ, D. (1957). *Arithmétique des Lois de Probabilité*. Mémorial des Sciences Mathématiques, **137**. Gauthier-Villars, Paris.

DYSON, F. J. (1953). Fourier transforms of distribution functions. *Canad. J. Math.*, **5**, 554-8.

ESSEEN, C. G. (1944). Fourier analysis of distribution functions. *Acta Math.*, **77**, 1-125.

FAINLEIB, A. S. (1968). A generalization of Esseen's inequality and its applications in probabilistic number theory. *Izv. Akad. Nauk SSSR, Ser. Mat.*, **32**, 859-79.

FELLER, W. (1971). *An Introduction to Probability Theory and its Applications*, vol. II. Wiley & Sons, New York (2nd edn).

FISHER, R. A. & DUGUÉ, D. (1968). Un résultat assez inattendu d'arithmétique des lois de probabilités. *C.R. Acad. Sci. Paris*, **227**, 1205-6.

FRYNTOV, A. E. (1974). The α-components of infinitely divisible laws. *Math. Physics and Functional Analysis*, **5**, 51-63. Akad. Nauk Ukr. SSR, Fiz. Tehn. Inst. Nizkih Temperatur, Kharkov.
[А. Е. Фрынтов. Об α-компонентах беэгранично делимых законов. Физико-техническ Институт Ниэких Температур.]

FRYNTOV, A. E. (1976). On the factorization of a countable number of Poisson laws. *Mat. Sb.*, **99** (**141**), 176-91. English transl., *Mat. USSR Sb.*, **28**, 153-67.

FRYNTOV, A. E. & ČISTJAKOV, G. P. (1977). On the membership of lattice probability laws in the class I_0. *Izv. Akad. Nauk SSSR, Ser. Mat.*, **41**, 1462-75. English transl., *Math. of the USSR, Izv.*, **41**, 441-52.

GNEDENKO, B. V. & KOLMOGOROV, A. N. (1954). *Limit Distributions for Sums of Independent Random Variables*. Addison-Wesley, Cambridge, Mass.

GOLDBERG, A. A. (1973). On a problem of Yu. V. Linnik. *Dokl. Akad. Nauk SSSR*, **211**, 31-4. English transl., *Soviet Mat. Dokl.*, **14**, 950-3.

GOLDBERG, A. A. & OSTROVSKII, I. V. (1967). An application of a theorem of W. K. Hayman to a problem in the theory of the decomposition of probability laws. *Ukr. Mat. Zh.*, **19**, 104-6.
[Гольдберг, А. А.-Островский, И. В. Применение теоремы У. К. Хеймана к одному вопросу теорий разложений вероятностних законов. Украинский Мат. Журнал, **19**, 104-6].

GOLDBERG, A. A. & OSTROVSKII, I. V. (1974). On the growth of entire ridge functions. *Math. Physics and Functional Analysis*, **5**, 3-10. Akad. Nauk Ukr. SSR, Fiz. Tehn. Inst. Nizkih Temperatur, Kharkov. (To be translated.)

GROSSWALD, E. (1976a). The Student t-distribution for odd degrees of freedom is infinitely divisible. *Ann. Probab.*, **4**, 680-3.

GROSSWALD, E. (1976b). The Student t-distribution of any degree of freedom is infinitely divisible. *Zeitschr. Wahrschr. verw. Gebiete*, **36**, 103-9.
GRUNKEMEIER, G. L. (1975). Decomposition of functions of bounded variation. *Ann. Probab.*, **3**, 329-37.
HARDY, G. H. (1963). *A Course of Pure Mathematics*. Cambridge Univ. Press (10th edn).
HEATHCOTE, C. R. & PITMAN, J. W. (1972). An inequality for characteristic functions. *Bull. Austral. Math. Soc.*, **16**, 1-9.
HILLE, E. (1959). *Analytic Function Theory*, vol. I. Ginn & Co., Boston & New York.
HILLE, E. (1962). *Analytic Function Theory*, vol. II. Ginn & Co., Boston and New York.
HILLE, E. (1964). *Analysis*. Blaisdell Publishing Co., New York.
HORN, R. A. (1972). On necessary and sufficient conditions for an infinitely divisible distribution to be Normal or Degenerate. *Zeitschr. Wahrsch. verw. Gebiete*, **21**, 179-89.
HUDSON, W. N. & TUCKER, H. G. (1975). Equivalence of infinitely divisible distributions. *Ann. Probab.*, **3**, 70-9.
IBRAGIMOV, I. A. (1956). On the composition of unimodal distributions. *Theory Probab. Applic.*, **1**, 255-60.
IBRAGIMOV, I. A. (1977). On determining an infinitely divisible distribution by its values on a half-line. *Theory Probab. Applic.*, **22**, 384-90.
IBRAGIMOV, I. A. & LINNIK, YU. V. (1971). *Independent and Stationary Sequences of Random Variables*. Walters & Noordhoff, Groningen. Russian original publ. by Nauka, Moscow, 1965.
ILINSKII, A. I. (1974). The indecomposable components of certain infinitely divisible laws. *Dokl. Akad. Nauk SSSR*, **215**, 529-31. English transl., *Soviet Mat. Dokl.*, **15**, 542-5.
ILINSKII, A. I. (1977). Indecomposable components of some infinitely divisible laws. *Teor. Funktsii Funktsional. Analis Priloz.*, vip. 27, Kharkov, 61-71.
[А. И. Ильинский. О неразложимых компонентах некогорых безгранично делимых законов. Теория Функций, Функциональний Анализ и их Приложения, 27. Харков, 61-71.]
ILINSKII, A. I. (1979). c-decomposability of characteristic functions. *Lithuanian Math. J.*, **18**, No. 4, 481-5.
ISMAIL, M. E. H. (1977). Bessel functions and the infinite divisibility of the Student t-distribution. *Ann. Probab.*, **5**, 582-5.
ISMAIL, M. E. H. & KELKER, D. (1976). The Bessel polynomials and the Student t-distribution. *SIAM J. Math. Analysis*, **7**, 82-91.
JESIAK, B. (1979). On analytic distribution functions and analytic properties of infinitely divisible distribution functions. *Teor. Veroyat. Primen.*, **24**, 825-31. English transl., *Theory Probab. Applic.*, **24**, 1979, 824-30.
JESIAK, B. (1981). On the unique determination of Lévy's spectral functions and parameters of infinitely divisible distributions. *Teor. Veroyat. Primen.*, **26**, 160-5. (Extension of this paper, preprint Leipzig.)
KAGAN, A. M., LINNIK, YU. V. & RAO, C. R. (1973). *Characterization Problems in Mathematical Statistics*. Wiley & Sons, New York. (Russian original publ. by Nauka, 1972.)
KAMININ, I. P. & OSTROVSKII, I. V. (1975). On the zeros of ridge functions. *Teor. Funktsii Funktsional. Analis Priloz.*, vip. 24, 41-50.
[И. П. Камынин-И. В. Островский. О нуляк целых хребтовых

функций. Теория Функций, Функциональний Анализ и их Приложения, **24**, 41-50.]
KANTER, M. (1976). On the unimodality of stable densities. *Ann. Probab.*, **4**, 1006-8.
KAWATA, T. (1969). On the inversion formula for the characteristic function. *Pacific J. Math.*, **31**, 81-5.
KAWATA, T. (1972). *Fourier Analysis in Probability Theory*. Academic Press, New York & London.
KEILSON, J. & STEUTEL, F. W. (1972). Families of infinitely divisible distributions under mixing and convolutions. *Ann. Math. Statist.*, **43**, 242-50.
KEILSON, J. & STEUTEL, F. W. (1974). Mixture of distributions, moment inequalities and measures of exponentiality and normality. *Ann. Probab.*, **2**, 112-30.
KELKER, D. (1971). Infinite divisibility and variance mixtures of the normal distribution. *Ann. Math. Statist.*, **42**, 802-8.
KUDINA, L. S. (1972). Indecomposable laws with a preassigned spectrum. *Teor. Funktsii Funktsional. Analis Priloz.*, **16**, 206-12. English transl. in *Selected Translations in Math. Statist. and Probab.*, **15** (1978), 69-76. Amer. Math. Soc.
KUDINA, L. S. (1973). The closure of the set of indecomposable distributions with a fixed spectrum. *Teor. Funktsii Funktsional. Analis Priloz.*, **16**, 51-6. (English transl. to be published.)
KUMAR, A. & SCHREIBER, B. M. (1978). Classification of subclasses of class L probability distributions. *Ann. Probab.*, **6**, 279-93.
LAHA, R. G. (1964). On the decomposition of a class of functions of bounded variation. *Canad. J. Math.*, **16**, 479-84.
LAHA, R. G. (1969). On an analytic decomposition of the Poisson law. *Trans. Amer. Math. Soc.*, **140**, 137-48.
LAPIN, A. I. (1947). On some properties of stable laws. (Dissertation.)
LAUE, G. (1980). Remarks on the relation between fractional moments and derivatives of characteristic functions. *J. Applied Probab.*, **17**, 456-66.
LEVIN, B. Ya. (1964). *Distribution of Zeros of Entire Functions*. Translations of Math. Monographs, vol. 5, Amer. Math. Soc., Providence, R.I.
LÉVY, P. (1937a). *Théorie de l'Addition des Variables Aléatoires*. Gauthier Villars, Paris.
LÉVY, P. (1937b). Sur les exponentielles de polynômes. *Ann. Scient. École Normale Supérieure* (3 Série), **73**, 231-92.
LÉVY, P. (1937c). L'arithmétique des lois de probabilité et les produits finis des lois de Poisson. *C.R. Acad. Sci. Paris*, **204**, 944-6.
LÉVY, P. (1952). Sur une classe de lois de probabilités indécomposables. *C.R. Acad. Sci. Paris*, **235**, 489-91. Also in P. Lévy's *Oeuvres*, vol. III, 518-19.
LÉVY, P. (1965). *Processus Stochastiques et Mouvement Brownien*. Gauthier Villars, Paris.
LÉVY, P. (1976). *Oeuvres*, vol. III, *Elements Aléatoires*. Gauthier Villars, Paris.
LEWIS, T. (1976). Probability functions which are proportional to characteristic functions and the infinite divisibility of the von Mises distribution. *Perspectives in Probability and Statistics* (publ. by Applied Probability Trust), 19-28.
LINNIK, Yu. V. (1964). *Decomposition of Probability Distributions*. Oliver & Boyd. (Russian original publ. in Moscow 1960 and transl. by S. J. Taylor.)
LINNIK, Yu. V. & OSTROVSKII, I. V. (1977). *Decomposition of Random*

Variables and Random Vectors. Translations of Math. Monographs, vol. 48, Amer. Math. Soc., Providence, R.I. (Russian original publ. 1972).

LINNIK, Yu. V. & SKITOVIČ, V. P. (1958). Again on the generalization of H. Cramér's theorem. *Vest. Leningr. Univ., Ser. Mat. Meh. Astron.*, **13**, 39-44.

[Ю. В. Линник-В. П. Скитович. Еще об обобщенях теоремы Г. Крамера. Вестник Ленинград. Унив., **13**, 39-44.]

LOÈVE, M. (1977). *Probability Theory*, vol. II. Springer, Berlin.

LUKACS, E. (1967). On the arithmetic properties of certain entire characteristic functions. *Proc. 5th Berkeley Symp. Math. Statist. and Probab.*, **2**, Part 1, 401-14.

LUKACS, E. (1968). Contribution to a problem of D. van Dantzig. *Teor. Veroyat. Primen.*, **13**, 116-27.

LUKACS, E. (1970). *Characteristic Functions.* Charles Griffin & Co., London (2nd edn).

LUKACS, E. (1975). *Stochastic Convergence.* Academic Press, New York (2nd edn).

MALOSHEVSKII, S. G. (1972). Infinite divisibility of a certain family of distributions. *Teor. Funktsii Funktsional. Analis Priloz.*, vip. **16**, 212-14.

[С. Г. Малощевский. Веэграничная делимость одного семейства распределений. Теория Функций, Функциональний Аналиэ и их Приложения, **16**, 212-14.]

MARCHAUD, A. (1927). Sur les dérivées et sur les différences des fonctions de variables réelles. *J. Math. Pures et Appl.*, sér. 9, t. 6, 337-425.

MARCINKIEWICZ, J. (1938). Sur les fonctions indépendantes, III. *Fund. Math.*, **31**, 86-102. Also contained in Marcinkiewicz' *Collected Papers* publ. by Panstwove Wydawnitvo Naukove, Warszawa (1964), 397-412.

MARKUSHEVIČ, A. I. (1965). *Theory of Functions of a Complex Variable.* Transl. by A. E. Silverman, Prentice-Hall, Engelwood Cliffs, N.J.

MASE, S. (1975). Decomposition of infinitely divisible characteristic functions with absolutely continuous Poisson spectral measure. *Ann. Statist. Math.*, **27**, 289-98.

MEDGYESSY, P. (1972). On the unimodality of discrete distributions. *Period. Math. Hungarica*, **2**, 245-57.

MESHALKIN, L. D. & ROGOZIN, B. A. (1963). An estimate of the distance between distribution functions by means of the closeness of their characteristic functions and its application to the central limit theorem. *Predel. Teoremi Teor. Veroyat., Izdat. Uzbek. SSR, Tashkent*, 49-55.

[Л. Д. Мещалкин-Б. А. Рогозин. Оценка расстояния между функциями распределения по близости их харктеристических функций и ее применение к центральной пределной теореме. Пределние Теореми Теория Вероятностей. Iэдат. А. Н. Уэбек. ССР Ташкент.]

MOTOO, M. (1955). Notes on a relation between the distribution functions and characteristic functions. *Ann. Inst. Statist. Math. Tokyo*, **6**, 191-5.

OBERHETTINGER, F. (1973). *Fourier Transforms of Distribution Functions and their Inverses.* Academic Press, New York & London.

OLSHIN, A. & SAVAGE, L. J. (1970). Generalized Unimodality. *J. Applied Probab.*, **7**, 21-34.

OSTROVSKII, I. V. (1963). Entire functions satisfying some special inequalities connected with the theory of characteristic functions of probability laws. *Kharkovsk. Univ. Mat. Obshestva*, **29**, 145-68. (English transl. in *Selected*

REFERENCES

Translations Math. Statist. and Probab., 7, 203-34. Amer. Math. Soc., Providence, R.I.)
OSTROVSKII, I. V. (1970a). On a class of characteristic functions. *Proc. Steklov Inst. Math.*, **111**, 195-207. (English transl. Amer. Math. Soc. (1972), 233-47.)
OSTROVSKII, I. V. (1970b). On some classes of infinitely divisible laws. *Izv. Akad. Nauk SSSR, Ser. Mat.*, **34**. (English transl. *Mat. USSR Izv.*, **4** (1970), 931-53.)
OSTROVSKII, I. V. & FLEKSER, P. M. (1973). Remarks on the argument of a characteristic function. *Mat. Fiz. Funktsional. Analis*, **4**, 13-14. (To be translated into English by Amer. Math. Soc.)
PADITZ, L. (1975). Eine ungleichmässige Abschätzung der Differenz zweier Verteilungsfunktionen. *Wiss. Z. Tech. Univ. Dresden*, **24**, 389-92.
PAULUSKAS, V. I. (1971). A smoothing inequality. *Litovsk Mat. Sb.*, **11**, 861-6. (In Russian, might be translated.)
PETROV, V. V. (1975). An inequality for the moments of a random variable. *Teor. Veroyat. Primen.*, **20**, 402-3. (English transl., *Theory Probab. Applic.*, **20**, 391-2.)
PÓLYA, G. (1949). Remarks on characteristic functions. *Proc. 4th Berkeley Symp. Math. Statist. & Probab.*, 115-23. Univ. California Press, Berkeley & Los Angeles.
PRAWITZ, H. (1973). Ungleichungen für den absoluten Betrag einer charakteristischen Funktion. *Skand. Aktuartidskr.*, 11-16.
PRAWITZ, H. (1974). Weitere Ungleichungen für den absoluten Betrag einer characteristischen Funktion. *Skand. Aktuartidskr.*, 21-8.
Prize Questions, *New Archief voor Wiskunde*, ser. 3, vol. 6 (1958), 28; ser. 7 (1959), 41; ser. 8 (1960), 42.
RAIKOV, D. A. (1939). On the composition of analytic distribution functions. *C.R. (Dokl.) Akad. Sci. USSR*, **23**, 511-14.
RAMACHANDRAN, B. (1962). Application of a theorem of Mamay's to a denumerable α-decomposition of the Poisson law. *Publ. Inst. Statist. Univ. Paris*, **13**, 13-19.
RAMACHANDRAN, B. (1967). *Advanced Theory of Characteristic Functions*. Statistical Publishing Society, Calcutta.
RAMACHANDRAN, B. (1969). On characteristic functions and moments. *Sankhyā*, **A**, **31**, 1-12.
RAMACHANDRAN, B. & RAO, C. R. (1968). Some results on characteristic functions and characterizations of the normal and generalized stable laws. *Sankhyā*, **A**, **30**, 125-40.
RIEDEL, M. (1975). On the one-sided tails of infinitely divisible distributions. *Math. Nachrichten*, **70**, 155-63.
RIEDEL, M. (1977). A new version of the central limit theorem. *Theory Probab. Applic.*, **22**, 183-4.
ROSSBERG, H. J. (1967). Der Zusammenhang zwischen einer ganzen charakteristischen Funktion einer verfeinerten Ordnung und ihrer Verteilungsfunktion. *Czechoslovak Math. J.*, **17** (**92**), 317-33.
ROSSBERG, H. J. (1974). On a problem of Kolmogorov concerning the normal distribution. *Theory Probab. Applic.*, **19**, 795-8.
ROSSBERG, H. J. (1979). Limit theorems for identically distributed summands, assuming the convergence of the distribution functions on a half axis. *Theory Probab. Applic.*, **24**, 693-711.

ROSSBERG, H. J. & JESIAK, B. (1978). On the unique determination of stable distribution functions. *Math. Nachrichten*, **82**, 297–301.

ROSSBERG, H. J. & SIEGEL, G. (1975). Continuation of convergence in the central limit theorem. *Theory Probab. Applic.*, **20**, 866–8.

ROSSBERG, H. J., JESIAK, B. & SIEGEL, G. (1981). *Continuation of Distribution Functions.* To be published in *Contributions to Probability*, Academic Press, New York.

RUEGG, A. (1970). A characterization of certain infinitely divisible laws. *Ann. Math. Statist.*, **41**, 1354–6.

SATO, K. (1978). Urbanik's class L_m of probability measures. *Ann. of Science, Kanazawa Univ.*, **15**, 1–10.

SATO, K. & YAMAZATO, M. (1978). On distribution functions of class L. *Zeitschr. Wahrsch. verw. Gebiete*, **43**, 273–308.

SENATOV, V. V. (1977). On some properties of metrics in the set of distribution functions. *Mat. Sb.*, **102** (144), 425–34.

SHAH, S. M. (1976). Entire functions whose Fourier transforms vanish outside a finite interval. *J. Math. Analysis Applic.*, **53**, 174–85.

SHAH, S. M. (1977). Exceptional values of entire characteristic functions. *Indian J. Pure and Appl. Math.*, **81**, 261–7.

SHIMIZU, R. (1970). On the domain of attraction of semi-stable distributions. *Ann. Inst. Statist. Math. Tokyo*, **22**, 245–55.

SHIMIZU, R. (1972). On the decomposition of stable characteristic functions. *Ann. Inst. Statist. Math. Tokyo*, **24**, 347–53.

SIEGEL, G. (1979). Uniqueness of symmetric distribution functions defined on bounded sets and its application. *Theory. Probab. Applic.*, **24**, 830–3.

STAUDTE, R. G. & TATA, M. N. (1970). Complex roots of real characteristic functions. *Proc. Amer. Math. Soc.*, **25**, 238–46.

STEUTEL, F. W. (1967). Note on the infinite divisibility of mixtures. *Ann. Math. Statist.*, **38**, 1303–5.

STEUTEL, F. W. (1968). A class of infinitely divisible mixtures. *Ann. Math. Statist.*, **39**, 1153–7.

STEUTEL, F. W. (1969). Note on completely monotone densities. *Ann. Math. Statist.*, **40**, 1130–1.

STEUTEL, F. W. (1970). *Preservation of Infinite Divisibility under Mixing.* Math. Center Tracts No. 33, Math. Centrum, Amsterdam.

STEUTEL, F. W. (1971). On the zeros of infinitely divisible densities. *Ann. Math. Statist.*, **42**, 812–15.

STEUTEL, F. W. (1973). Some recent results in infinite divisibility. *Stochastic Processes and Applic.*, **1**, 125–43.

STEUTEL, F. W. (1974). On the tails of infinitely divisible distributions. *Zeitschr. Wahrsch. verw. Gebiete*, **28**, 273–6.

STEUTEL, F. W. & VAN HARN, K. (1979). Discrete analogues of self-decomposability and stability. *Ann. Probab.*, **7**, 893–9.

STUDNEV, Yu. P. (1967). Some generalizations of limit theorems in probability theory. *Theory Probab. Applic.*, **12**, 688–72.

STUDNEV, Yu. P. (1970). The theory of infinitely divisible laws in the class B. *Theory of Probability and Mathematical Statistics*, vol. 2, 187–96. (Russian original 1970.)

SZÁSZ, D. O. H. (1973). On a rolling characteristic function. *Period. Math. Hungarica*, **3**, 13–17.

SZEGÖ, G. (1959). *Orthogonal Polynomials.* Amer. Math. Soc., Colloquium Publications, vol. 23.

TEUGELS, J. L. (1971). Probability density functions which are their own characteristic functions. *Bull. Sci. Math. Belg.*, **23**, 236-72.
THOMPSON, J. (1975). A note on the Lévy distance. *J. Appl. Probab.*, **12**, 412-14.
THORIN, O. (1977a). On the infinite divisibility of the Pareto distribution. *Skand. Aktuartidskr.*, 31-40.
THORIN, O. (1977b). On the infinite divisibility of the lognormal distribution. *Skand. Aktuartidskr.*, 121-48.
TUPICINA, V. M. (1972). On the arithmetic of ridge functions. *Teor. Funktsii Funktsional. Analis Priloz.*, vip. 15, 142-51. (To be translated.)
URBANIK, K. (1968). A representation of self-decomposable distributions. *Bull. Acad. Polon. Sci., Sect. Math. Astr. Phys.*, **16**, 209-14.
URBANIK, K. (1973). Limit laws for sequences of normed sums satisfying some stability conditions. *J. Multivariate Analysis*, **3**, 225-37.
URBANIK, K. (1975). *Extreme Point Methods in Probability Theory.* Lecture Notes in Math., No. 472, 169-94. Springer, Berlin-Heidelberg.
VALIRON, G. (1949). *Lectures on the General Theory of Integral Functions.* Chelsea Publishing Co., New York. (French edn, 1923.)
VALIRON, G. (1959). Fonctions entières d'ordre fini et fonctions méromorphes. *L'Enseignement Mathématique No. 8, Genève.*
VAN DANTZIG, D., see Prize Questions.
VAN HARN, K. (1978). Classifying infinitely divisible distributions by functional equations. Math. Centrum, Amsterdam.
VINNICKI, B. V. (1975). On a property of infinitely divisible characteristic functions of probability laws. *Izv. Vishiy Učeb. Zaved. Mat.*, No. 4 (155), 95-7.
WIDDER, D. V. (1946). *The Laplace Transform.* Princeton Univ. Press.
WINTNER, A. (1956). Cauchy's stable distribution and an explicit formula of Mellin. *Amer. J. Math.*, **78**, 819-61.
WOLFE, S. J. (1971a). On the unimodality of L-functions. *Ann. Math. Statist.*, **42**, 912-18.
WOLFE, S. J. (1971b). On the continuity properties of L-functions. *Ann. Math. Statist.*, **42**, 2064-73.
WOLFE, S. J. (1973). On the local behaviour of characteristic functions. *Ann. Probab.*, **1**, 862-6.
WOLFE, S. J. (1975). *On Moments of Probability Distributions.* Lecture Notes in Math., No. 457, 306-16. Springer, Berlin-Heidelberg.
YAKOVLEVA, N. I. (1972). The growth of characteristic functions of probability laws. *Teor. Funktsii Funktsional. Analis Priloz.*, vip. 15, 43-9.
[Юаковьлева, Н. И. О росте целых характеристических функций вероятностных законов. Теория Функций Функционалний Анализ и их Приложения, **15**, 43-9. Will be translated.]
YAKOVLEVA, N. I. (1976). On the growth of characteristic functions of probability laws. *Vop. Mat. Fiz. Funktsional. Analis, Fiziko Tehn. Inst. Niznih Temperatur, Akad. Nauk USSR*, 43-54.
[Юаковьлева, Н. И. О росте целых характеристических функций вероятностных законов. Вопроэи Мат. Физики и Функционалний Анализ, А. Н. YSSR, 43-54. Will be translated.]
YAKOVLEVA, N. I. (1978). Decomposition of certain infinitely divisible distribution functions and composition of functions of bounded variation. *Ukr. Mat. Zh.*, **30**, 273-8. (In Russian; will be translated.)
YAMAZATO, M. (1975). Some results on infinitely divisible distributions of

class L with applications to branching processes. *Science Reports Tokyo Kyoiku Daigaku*, sect. A, **13**, No. 362, 133-9.

ZIMOGLJAD, V. V. (1969). On a class of entire functions which satisfy the ridge property. *Vop. Mat. Fiz. Funktsional. Analis, Trudi Fiz. Tehn. Inst. Nizkih Temperatur Ukr. SSR*, **1**, 172-90.

[Эимогляд, В. В. Об одном классе целых функций, обладающих свойством "хребта". Труды Физико-технического Института Ниэких Температур. А. N. YSSR, **1**, 172-90.

ZOLOTAREV, V. M. & SENATOV, V. V. (1975). Two-sided estimates of the laws of the L-class. *Litovsk Mat. Sb.*, **3**, 123-40. (In Russian.)

ZOLOTAREV, V. M. (1965). On the closeness of distributions of two sums of independent random variables. *Theory Probab. Applic.*, **10**, 472-8.

ZOLOTAREV, V. M. (1967). A sharpening of the inequality of Berry-Esseen. *Zeitschr. Wahrsch. verw. Gebiete*, **8**, 332-42.

ZOLOTAREV, V. M. (1970). Some new inequalities in probability connected with Lévy's metric. *Soviet Mat. Dokl.*, **11**, 231-4.

ZOLOTAREV, V. M. (1971). Estimates of the difference between distributions in the Lévy metric. *Proc. Steklov Inst. Math.*, **112**, 232-40.

ZOLOTAREV, V. M. (1976). Metric distances in spaces of random variables and their distributions. *Mat. Sb.*, **101** (143), 416-54.

ZOLOTAREV, V. M. (1977). Ideal metrics in the problem of approximating distributions of sums of random variables. *Theory Probab. Applic.*, **22**, 433-49.

ZOLOTAREV, V. M. (1978). Some remarks on the paper by Rossberg-Jesiak (1978). *Math. Nachrichten*, **82**, 301-5.

ZOLOTAREV, V. M. & SENATOV, V. V. (1975). Two-sided estimates of the Lévy metric. *Theory Probab. Applic.*, **20**, 234-45.

Index

Abramov, V. A., 43
Alf, C., 66
α-decomposition, 112
—, finite, 112
—, denumerable, 112
analytic c.f., 14, 112
— —, continuation of, 76
— d.f., 76, 122
arcsine d., 73
Askey, R., 52

Barndorff-Nielsen, O., 68
Bartholomew, D. J., 119
Berman, S. M., 11
Beta d., 72
between two metrics, 135
Boas, R. P., 69, 110, 124
Bochner, S., 4
boundary c.f., 116
bounded variation, 150

canonical representation, 12, 13, 45
   Lévy's — —, 12
   Kolmogorov — —, 93
Cauchy d., 71
characteristic function (c.f.), 2, 3, 11
— —, indecomposable, 71
— —, decomposable, 71
— —, analytic, 76
— —, entire, 80
Chung, K. L., 50
Cistjakov, G. P., 101, 104, 113, 156
class L_m, 48
— L, 47, 127
— N, 85
— $\wedge$, 85
— $\mathscr{L}$, 100, 114
— $\mathfrak{D}$, 92

— $\mathfrak{D}_s$, 93
— $\mathfrak{D}_E$, 96
— I_0, 99, 103, 104
— I_0^α, 114
— $\mathscr{P}$, 98
closeness of d.f., 31
comparison of metrics, 131
completely monotone, 67
continuation of d.f., 127
continuity point, 1
continuity theorem, 3, 31
convergence, weak, 32
convolution theorem, 3
convolution of unimodal d.f., 50
Cuppens, R., 104
Cramér, H., 150

Dantzig, D. van, 92
decomposable c.f., 14, 70
decomposition, 112, 113
—, of f. of bounded variation, 153
Dewess, M., 85, 166
Dharmadikari, S. W., 53
discontinuity point, 13
distance between c.f., 135
distance between d.f., 131
distance definition, 31
distribution function (d.f.), 1
— —, 1
— —, 1
— —, 14
— —, 12
Dugué, D., 92, 113, 140
Dyson, F. W., 136

entire c.f., 80
entire d.f., 86, 125
entire function, 164

* The following abbreviations are used in the index: c.f. = characteristic function, a.c.f. = analytic characteristic function, d. = distribution, d.f. = distribution function, f. = function, i.d. = infinitely divisible, i.d.d. = infinitely divisible distribution.

Esseen, C. G., 33
estimate of distance, 135
expansion of c.f., 21, 23

factorization, 14, 65
factor, indecomposable, 14
Fainleib, A. S., 43
family $\mathscr{G}_2$, 86
finite d., 69
Fisher, R. A., 92
Flekser, P. M., 11
Fourier-Stieltjes transform, 2, 150
fractional calculus, 25
— derivative, 25
— moment, 25
Fryntov, A. E., 101, 103, 115
functions of class B, 151
— — —, B_1, B_2, B_3, 155
— — —, $B_4, \mathfrak{B}$, 156
— — —, B_0, 153

geometric d., 73
Gnedenko, B. V., 50
Goldberg, A. A., 16, 100, 148
Grosswald, E., 67

Hadamard, 167
Halgreen, C., 68
Hardy, G. H., 22
Heathcote, C. R., 7
Hermite polynomial, 87
Hermitian property, 77
Hille, E., 150, 165, 167, 168
Horn, R. A., 70
Hudson, W. N., 67
Hurwitz, A., 143
hyperbolic d., 68

Ibragimov, I. A., 53, 54, 57, 105, 110
Ilinskii, A. J., 48, 73, 74
indecomposable d.f., 71
— ridge function, 145, 146
— factor, 99
infinitely divisible d., 65, 105
— — c.f., 65, 105
inverse Gaussian d., 68
inversion theorem, 3
Ismail, M. E. H., 67

Jensen's inequality, 6
Jesiak, B., 110, 125, 126, 127
Jogdeo, K., 53

Kagan, A., 45, 141

Kaminin, I. P., 148
Kanter, M., 57
Kawata, T., 3, 27, 125
Keilson, J., 119
Kelker, D., 67
Khinchine, A. Ya., 49
Kolmogorov, A. N., 31, 50, 93, 105, 127
Kudina, L. S., 74, 75
Kumar, A., 98

Laha, R. G., 114, 154
Lapin, A. I., 50
Laplace d., 122
Laue, G., 30
lattice d., 101
L-class, 14, 47, 58, 62
$\mathscr{L}$-class, 111, 114
Levin, B. Ya., 82
Lévy, P., 11, 31, 38, 73, 101, 102
Lévy-Khinchine representation, 13
Lévy canonical representation, 12, 58, 105
Lewis, T., 68
limit theorems, 128
Linnik, Yu. V., 45, 57, 100, 101, 113, 114, 141, 154
Loève, M., 11, 34
log-normal d., 68
Lukacs, E., 3, 4, 7, 50, 76, 107, 112, 113, 138, 142, 147, 153

Maloshevski, S. G., 68
Marchaud, A., 25
Marcinkiewicz, J., 15, 110, 116, 148
Markushevič, A. I., 110, 141, 146, 165, 167
Mase, S., 74
maximum modulus, 86
Medgyessy, P., 53
Meshalkin, L. D., 43
metric(s), 131
—, uniform, 131
—, Lévy, 131
—, Kolmogorov, 131
—, stronger, 131
—, weaker, 131
—, equivalent, 131
—, monotone, 135
— in space of c.f., 135
metrics, ideal, 135
metric space, 31

INDEX

Mises d., 68
mixtures, 119
— of d.f., 119
— of exponential d., 119
— of i.d., 119
— of exponential frequencies, 121
— of gamma d., 121
moment(s), 65
—, algebraic, 2
—, absolute, 2
—, of positive order, 17
Motoo, M., 43

non-negative definite, 4
normal d., 105, 113

Oberhettinger, F., 2
O'Connor, T., 66
Olshen, A., 52
one-sided d., 82
operators T_1, T_2, 87
order, 81, 82, 164
— of ridge function, 148
Ostrovskii, I. V., 11, 96, 100, 101, 113, 114, 115, 141, 146, 148

Paditz, J. L., 43
Pareto d., 68
Paulauskas, V. I., 43
Petrov, V. V., 2
Pitman, J. W., 7
Pólya, G., 4, 5, 51, 70, 161
Pólya's theorem, 70
Poisson spectrum, 100
— type d., 101, 103
Prawitz, H., 7
proximate order, 82
— type, 82

Raikov, D. A., 124, 150, 160, 163
Ramachandran, B., 46, 114, 161
Rao, C. R., 45, 46, 141
rectangular d., 40
relatively compact, 130
restricted convergence, 128
ridge function, 15, 138, 139
— —, standardized, 138, 142
— —, zeros of, 139, 142
— —, factorization of, 143
— —, component of, 143
— —, factor of, 143

Riedel, M., 85, 110, 129, 166
Rogozin, B. A., 43
Rossberg, H. J., 84, 85, 105, 110, 127, 129, 166
Ruegg, A., 69

Sato, K., 48
Savage, L. J., 52
scale invariant, 33
Schreiber, B. M., 48
Schwarz's reflection principle, 167
self-decomposable d.f., 14, 47, 48, 58
— — c.f., 14, 47, 48, 58
self-reciprocal, 93
semi-stable d., 45
Senatov, V. V., 31, 134
Shah, S. M., 99
Shimizu, R., 8, 46, 71
Siegel, G., 110, 127
singular point, 76, 77
singularity, 76, 77
Skitovič, V. P., 154
Smithies, F., 124
spectral function, 13, 65
spectrum, 73
stable d.f., 13
stable d., 45
— —, generalization of, 45
— —, extension of, 45
stable c.f., 13, 45
Staudte, R. G., 68
Steutel, F. W., 66, 70, 110
strip of regularity, 15, 77
strong unimodality, 53, 54, 56
Student's t-distribution, 67, 122
Studnev, Yu. P., 151
symmetric d.f., 23
— difference, 23
— stable d.f., 70, 122
symmetrized d.f., 23
Szász, D. A., 11, 137
Szegö, G., 87

tail, 19, 68
tail behaviour, 19, 110
Tata, M. N., 68
Teugels, J. L., 98
Thorin, O., 68, 69
translation invariant, 33
Tucker, H., 67
Tupicina, V. M., 146
type, 82, 164

ultimately positive, 157
unimodal, 49
unimodality, 49
uniqueness theorem, 3
Urbanik, K., 48

Valiron, G., 165
van Harn, K., 67
van Dantzig, D., 92
van Dantzig class, 92
variance mixture, 122
— — of normal d., 122
vertex, 50
Vinnickii, B. V., 86

weak convergence, 32
Weierstrass primary factors, 167
Widder, D. V., 157
Wintner, A., 58
Wolfe, S., 28, 30, 47

Yakovleva, N. I., 86, 163
Yamazato, M., 58, 59, 62

Zimogljad, V. V., 148
Zolotarev, V. M., 38, 43, 127, 135, 137